Fillers

Edited by Emmanuel Flores Huicochea

Published in London, United Kingdom

IntechOpen

Supporting open minds since 2005

Fillers
http://dx.doi.org/10.5772/intechopen.87875
Edited by Emmanuel Flores Huicochea

Scientific Contributor: Xiaojian Zhou

Contributors
Xiaojian Zhou, Guanben Du, Long Cao, Mattia Bartoli, Mauro Giorcelli, Pravin Jagdale, Massimo Rovere, Magdalena Lipińska, Klaudia Klara Toczek, Magdalena Stefaniak, Abbas Hasan Faris, Giovani Rissi

First published in London, United Kingdom, 2021 by IntechOpen
IntechOpen is the global imprint of INTECHOPEN LIMITED, registered in England and Wales, registration number: 11086078, 5 Princes Gate Court, London, SW7 2QJ, United Kingdom
Printed in Croatia

British Library Cataloguing-in-Publication Data
A catalogue record for this book is available from the British Library

Additional hard and PDF copies can be obtained from orders@intechopen.com

Fillers
Edited by Emmanuel Flores Huicochea
p. cm.
Print ISBN 978-1-83962-435-3
Online ISBN 978-1-83962-436-0
eBook (PDF) ISBN 978-1-83962-437-7

Meet the editor

Dr. Emmanuel Flores Huicochea obtained his bachelor's degree from Universidad Autónoma del Estado de Morelos and later obtained a Ph.D. degree from the Chemistry Faculty, Universidad Nacional Autónoma de México in Engineering in 2013. He has been a full-time researcher and associate professor of postgraduate studies in the Centro de Desarrollo de Productos Bióticos belonging to Instituto Politécnico Nacional, México. Additionally he has contributed to research projects and has published research papers and book chapters. His investigation topics are related to starch modification and use to produce bioplastics added with lignocellulosic to use as a barrier or disposable container.

Contents

Preface XI

Chapter 1 1
Fillers in Wood Adhesives
by Abbas Hasan Faris

Chapter 2 19
Wood Adhesive Fillers Used during the Manufacture of Wood Panel Products
by Long Cao, Xiaojian Zhou and Guanben Du

Chapter 3 29
Fillers for Packaging Applications
by Giovani Otavio Rissi

Chapter 4 43
Towards Traditional Carbon Fillers: Biochar-Based Reinforced Plastic
by Mattia Bartoli, Mauro Giorcelli, Pravin Jagdale and Massimo Rovere

Chapter 5 57
POSS Fillers as a Factor Influencing on Viscoelastic Properties, Crystallization, and Thermo-Oxidative Degradation of Poly(Lactic Acid)-Epoxidized Natural Rubber PLA/ENR Blend
by Magdalena Lipińska, Klaudia Toczek and Magdalena Stefaniak

Preface

It is my utmost pleasure to write the preface for this book entitled "Fillers", which is written by different researchers in the world. Fillers, in general, are substances that can be added to various polymer systems to reduce the cost or improve their properties (mechanical properties, visual performance, and other physical-chemistry performances). It can be either organic or inorganic, and added into the main matrix in different ways as a solid, liquid, or gas with different dimension scales, including macro, micro, even nanoscale.

How they can be used efficiently in wood adhesive, plastic, poly (lactic acid, PLA)/ rubber composite, and packaging? The aim of this book is to present the most updated information on the fillers in different application fields. In addition, the most recent research achievements will also be covered.

I fully believe that the authors' excellent work will have significant academic value and a far-reaching impact, with extensive applicability and practical significance to the polymers industry.

Dr. Xiaojian Zhou
Southwest Forestry University,
Panlong District, Kunming, Yunnan Province, China

Chapter 1

Fillers in Wood Adhesives

Abbas Hasan Faris

Abstract

The introduction of a second component to polymers has been presented; this component is often used to modify the characteristics of the products and to acquire new polymer materials with improved properties. Composite materials have a pivotal role in industries that are now considered the most progressive worldwide. At present, synthetic adhesives based on formaldehyde such as phenol-formaldehyde (PF), urea formaldehyde (UF), and melamine formaldehyde (MF) are predominantly used for wood composite production, and these adhesives are commonly used in the wood panel industry. These adhesives have some advantages and disadvantages. The use of PF adhesives is as important as UF adhesives in the wood panel industry. However, their application is still limited because of its brittleness, brown color, high curing temperature, long curing time, and toxicity due to liberation of phenol and formaldehyde. A variety of methods have been used to improve the performance of UF and PF adhesives as well as to expand their use. These methods are widely used in the industry; they include the simple addition of fillers. Moreover, the addition of fillers could reduce shrinkage and alleviate the stress on the glue line, which improves the hardness and durability.

Keywords: bonding strength, PF, wheat flour, tannin adhesive, organoclay

1. Introduction

The introduction of a second component to polymers has been presented; this component is often used to modify the characteristics of the products and to acquire new polymer materials with improved properties. Composite materials have a pivotal role in industries that are now considered the most progressive worldwide. Generally, the term "composite" is given to materials made of more than one component. Polymer composites are a mixture of polymers with inorganic or organic additives that have certain geometric forms (spheres, flakes, fibers, and particulates). A wide range of polymer characteristics can be improved by composite technologies, such as their mechanical, thermal, barrier, flame retardant, magnetic, optical, and electrical properties and durability, chemical stability, corrosion resistance, and biodegradability [1–4].

Wood adhesives now represent a vital aspect of the wood-based panel industry, and synthetic condensation resins have become widely used. The synthetic resins used can be classified into four varieties, and all these adhesives are based on formaldehyde. These resins are urea formaldehyde (UF), phenol-formaldehyde (PF), resorcinol formaldehyde (RF), and melamine formaldehyde (MF). The PF resin is used in wood adhesives, thermal insulation materials, coatings, molding compounds, and other applications. The high stability of the C-C linkages between the aromatic ring and methylene bridge and the resistance to hydrolysis make it a favorite resin for

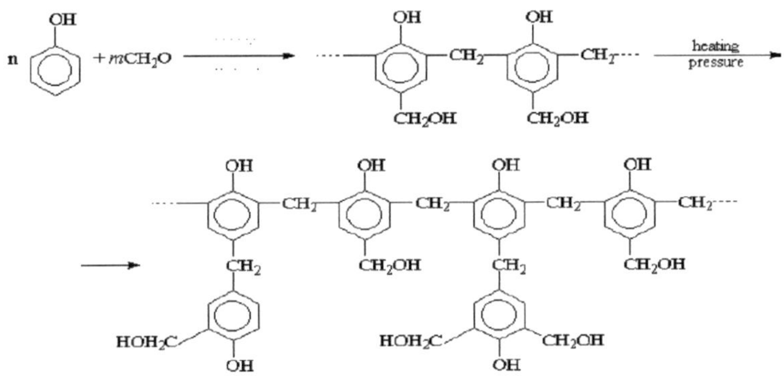

Figure 1.
Synthesis of phenol-formaldehyde resins by polycondensation of phenols, in excess of formaldehyde developed by Baekeland.

glue lines as well as boards, such as weatherproof plywood, oriented strand boards (OSB), medium-density fiber boards (MDF), or particleboards for use under exterior weather conditions. The global demand for PF resin (resol, novolac, and others) continues to increase; recent reports expect this demand to grow at an annual rate of 5.7% from 2014 to 2019 by $19.31 billion. This increased demand is associated with the increased construction in developing regions [5]. The resol phenolic resin accounted for more than 75% of the global market in 2013. This market is expected to grow at a healthy rate of 4.1% until 2019. The PF resin is obtained via condensation reactions between phenol and formaldehyde in the presence of an acid or a base to produce novolac or resol, respectively. Resol-type PF resins (**Figure 1**) [6] have been widely used for decades to manufacture wood adhesives due to their high performance in terms of the mechanical and thermal properties as well as its water resistance.

The concern regarding crude oil supplies and the toxicity of the raw materials extracted from the fossil fuels has grown as they can cause climate change. According to findings conducted by the National Cancer Institute (NCI) on a group of workers who were exposed continuously to formaldehyde, a high possibility of cancer was diagnosed, prompting the World Health Organization to the reclassification of formaldehyde to group I. Wood-based panel industry relies heavily on this material, making it difficult to impose an outright ban on the use of formaldehyde.

Recently, the development of natural or green-based wood adhesives as successful substitutes for synthetic resins has become of interest because of the unwarranted increase in the prices of fossil fuels as well as environmental and health concerns. Lignin and tannin are materials rich in phenolic compounds; both were found to be successful alternatives for phenol in the manufacturing of bio-based PF resins.

2. Fillers

Fillers are substances that can be added to various polymer systems to reduce the cost or improve their properties [7, 8]. This material can be added as a solid, liquid, or gas. For example, the use of a minimal percentage of clay loading can lead to significant improvements in the mechanical and thermal properties [7, 8]. In general, the fillers used to modify the properties of polymers can be classified into two categories: inert fillers and active fillers. Inert fillers come from inorganic mineral powders, such as kaolin, diammonium phosphate, porcelain clay (which is frequently used), sodium silicate, and magnesium oxide. These fillers are used to reduce cost. These mostly hydrophilic materials can be dispersed in adhesives. Active fillers are organic compounds, which swell when dissolved in a solution.

These fillers include the cellulose-type fillers (wood powder, bark powder, etc.), protein-type fillers (soy protein, blood powder, etc.), and starch-type fillers (wheat flour, etc.). These materials have the ability to react with resins.

Clays are one group of the additives that have been widely used in the preparation of polymeric composite materials. Recently, increasing attention has been given to the development of polymer/clay nanocomposites because of their dramatically improved properties as compared with conventional fillers after the addition of very low portions of the filler [9]. Clay minerals are used in soil science and sedimentology to refer to particles formed by the combination of tetrahedral and octahedral aluminosilicate ($[SiO_4]^{4-}$ and $[AlO_4]^{5-}$) layers that are 1 nm thick with 200–300 nm lateral dimensions, as shown in **Figure 2** [10]. Clay minerals are important materials that are mainly hydrated aluminosilicates with neutral or negative charges [11]. The internal and external cations can be changed by inorganic or organic cations [10, 12].

The hydrophilic nature of clays makes them poorly suited for mixing with most hydrophobic polymer materials [13]. In addition, the poor physical interaction between the organic components in polymeric materials and the inorganic components in clay leads to their separation and the formation of discrete phases. Therefore their mixtures have poor properties, and this incompatibility prevents the dispersion of clay layers within polymer matrix, thereby leading to weak interfacial interactions [14–16]. Moreover, electrostatic forces tightly link the clay platelets with each other. Only some hydrophilic polymers are miscible with layers of clay; these materials are used in the preparation of polymer-clay composites such as poly(vinyl alcohol) and poly(ethylene oxide) [8].

For these reasons, clay requires pretreatment before it is used as filler in polymer composites. These piles of clay platelets have dimensions much larger than 1 nm. Therefore, untreated clay would be ineffective during use because most of the clay would be trapped inside and unable to react with the polymer matrix. Generally, the intercalation of clay with different organic species is the basic condition for compatible composite materials to ensure the interaction between the clay surfaces and the polymeric components. The intercalation or surface modifications decrease the surface energy of clay layers and match their surface polarity with polymer polarity. The low surface energy of organoclays is more compatible with organic polymers and can intercalate with the interlayer space under specific experimental conditions.

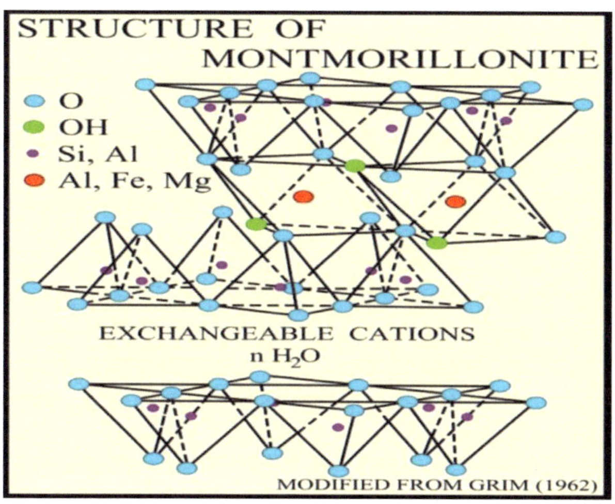

Figure 2.
Montmorillonite structure.

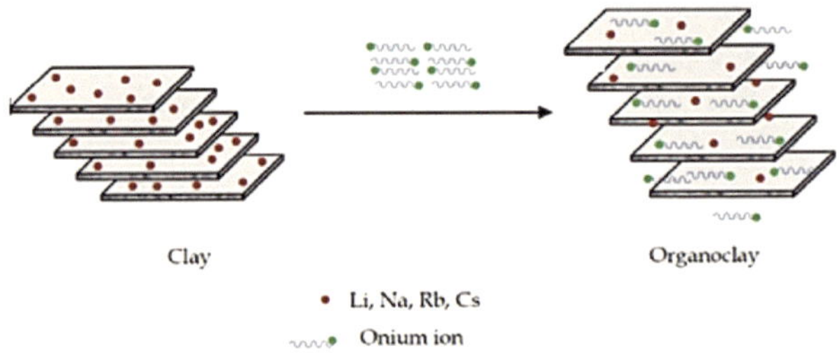

Figure 3.
Scheme of the modification of clay layers by organic onium cations.

This spacing of organoclays is affected by a variety of factors, including the chemical structure of the surfactant, the degree of cation exchange, and the silicate layer thickness [17]. Organically modified clays have been extensively studied in various practical applications in the field of organic-inorganic hybrids, composites, and nanoscale composites [15]. The surface modification of clay layers can be performed through an ion exchange process via the replacement of cations, such as sodium and calcium, in the interlayer space by ammonium or phosphonium surfactants, which usually include benzyl groups and short aliphatic chains [18]. In addition to modifying the surface and increasing the hydrophobicity of clay layers, the introduction of alkylphosphonium ($R_4P^+X^-$) and alkylammonium ($R_4N^+X^-$) cations into the clay layers increases the distances and spacing of the clay layers, which facilitates the intercalation of the polymer chain during the preparation of nanocomposites [19]. Moreover, the cations ($R_4N^+X^-$) and ($R_4P^+X^-$) can provide the necessary functional groups that have the ability to interact with the polymer chains or to initiate polymerization processes, thereby increasing the interfacial interactions, as shown in **Figure 3**.

3. Modification of wood adhesives using fillers

At present, synthetic adhesives based on formaldehyde such as PF, UF, and MF are predominantly used for wood composite production. These adhesives have some advantages and disadvantages. For example, UF adhesives have advantages such as the lack of color in the cured polymer, low price, good mechanical properties, and so on. Consequently, these adhesives are commonly used in the wood panel industry. However, UF resin can only be used as an interior adhesive because of its poor water resistance. The use of PF adhesives is as important as UF adhesives in the wood panel industry. The good water resistance of PF adhesives allows them to be used under more stringent conditions than UF adhesives; thus, PF adhesives can be classified as exterior adhesives. However, their application is still limited because of its brittleness, brown color, high curing temperature, long curing time, and toxicity due to liberation of phenol and formaldehyde [20]. A variety of methods have been used to improve the performance of UF and PF adhesives as well as to expand their use. These methods are widely used in the industry; they include the simple addition of fillers. A variety of resin properties can significantly improve their performance as wood adhesives by the addition of fillers, which are most important for their lower cost. The mechanical performance of the resin is also improved in most cases. Moreover, the addition of fillers could reduce shrinkage and alleviate the stress on the glue line, which improves the hardness and durability. The flow of the adhesive and its smooth flow on the wood surface will be improved because of the penetration of water into the glue in the presence of a filler. The moderate permeability of the glue between the wood components has good mechanical effects by physically interlocking the wood surface with the curing agents; thus, the bonding strength between the wood products is improved.

Wheat flour is an activated filler, which is most commonly used in factories for the production of UF resins. The addition of approximately 10 to 20% of the wheat flour to the total solid resin can increase the weather resistance of the resin because this filler can reduce the shrinkage of the glue line [21]. According to Ma [22], the use of oxidized starch as a filler with UF resin at proportions of 5–10% will decrease the free formaldehyde emission from 3–7% to 0.2–0.5%. The hydroxyl group reacts with starch molecule chains and liberates formaldehyde to form acetals, thereby reducing the free formaldehyde emission. Compared with wheat flour, oxidized starch as a filler can further increase the bonding strength and water resistance of UF resins [23].

Mosiewicki et al. [24] studied the mechanical and thermal behavior of a composite formulated from a natural quebracho tannin adhesive with pine wood flour as well as the effect of moisture on its adhesive properties. The results showed that the mechanical and thermal properties of composite materials of wood flour/tannin adhesives have sufficient values for use in some industrial applications in terms of its higher stiffness, which is an important requirement. However, this composite material can be used in the dried state alone and must be avoided in humid atmospheres.

Kargarfard and Jahan-Latibari [25] used recycled polyethylene with UF to reduce swelling fish and water absorption during particleboard production, thereby improving its performance in terms of weather resistance. In this work, different amounts of recycled polyethylene (5%, 10%, and 15%) were mixed with the lowest amount of UF resin (4%). Their results revealed that 5% of recycled polyethylene in the surface layer could improve the weather resistance of particleboards. The modulus of rupture (MOR) and modulus of elasticity (MOE) properties were marginally increased when the percentage of recycled polyethylene was increased to 5% in the surface layer. However, when 15% recycled polyethylene was added to the surface layer, the increased MOR was almost doubled, and the MOE was increased by approximately 50% as compared with the control boards.

Du et al. [26] used the synergistic effect of mineral filler (MgO and SiO_2) with UF adhesives for the production of plywood. Plywood has better performance with mineral fillers as compared with the absence of filler or with wheat flour fillers. These results can be attributed to two reasons. (i) The (MgO and SiO_2) filler prevents the glue from excessive permeation into the wood substrate and reduces the internal stress caused by the shrinkage. (ii) The oxygen from silicate and magnesium may be linked with the hydroxymethyl groups of the UF adhesive, thereby improving the durability and weather resistance of the resins.

Another newly developed type of mineral fillers are the nanomaterials, which differ physically and chemically from the normal fillers in terms of their thermal, light, radiation, and mechanical characteristics, among others [27]. Polymer nanocomposites have attracted considerable attention from researchers, especially in research related to polymer-layered silicate (PLS). This interest is attributed to the fact that the addition of small amounts of these materials often significantly improves the mechanical and thermal properties of polymers.

Some researchers have tried to use the nanomaterial fillers in wood adhesive systems, such as nano-SiO_2, nano-$CaCO_3$, and nano-montmorillonite (MMT) [28–31]. Their results showed that the adhesive bonding strength, gel time, and free formaldehyde emission were affected by the addition of nanomaterials. In most cases, the emission of free formaldehydes was significantly reduced.

The exfoliated nano-MMT was used by Zhang et al. [32] as a nanomineral filler with polymeric MDI to modify degraded soybean protein (DSP) and improve its water resistance and technical applicability. The DSP adhesives with MDI as a cross-linking agent showed improved water resistance and bond strength, but their shelf life is very short. After the modification of MDI-modified DSP by nano-MMT, the produced adhesive had a much longer shelf life but slightly lower bonding strength.

Various PLSs have been successfully used with various types of resins, such as epoxy [33], polyacrylic ester, polyurethane [34], and even PF [35]. However, research on the feasibility and the mechanism of wood adhesive modification with MMT is rather limited.

4. Preparation of water-resistant wood adhesives

Glyoxalated lignin-tannin (GLT) adhesives are good candidates for the replacement of formaldehyde-based adhesives because of health and environmental concerns. Although glyoxalated lignin-tannin resins are low cost and have environment-friendly properties, these types of adhesives do not meet the fundamental required bonding strength and water resistance [36]. In addition, poor water-resistant property has limited their application. Ease of hydrolysis makes this adhesive useful for interior applications only. Consequently, appropriate measure must be taken to overcome the problem and to improve the water resistance of adhesives. In this work, we tried to develop and improve the properties of glyoxalated lignin-tannin resin to meet the fundamental requirement of bonding strength and water resistance.

4.1 Preparation of green wood adhesives using modified tannin and hyperbranched poly(amine-ester)

Due to the unique structure, chemical and physical characteristics, and various industrial applications, hyperbranched polymers (HBPs) have become the focus of considerable interest to chemists, biochemists, biologists, and biomedical experts [37]. This is due to high solubility, reduction of melt and solution viscosity, and abundance of functionalities as a result of the large number of reactive terminal groups within a molecule, nearly spherical molecular shape and the absence of chain entanglement [38]. These characteristics make dendrimers and hyperbranched polymers (HBPs) applicable as drug delivery agents, catalysis, Mitchell mimics, and nanoscale building blocks to artificial cells and coatings [39]. Both dendrimers and hyperbranched polymers are three dimensional and highly branched macromolecules. In this work, oligomeric precursors of poly(amine-ester) were synthesized, and then they were used to modify the tannin to improve the water resistance and mechanical properties of glyoxalated lignin-tannin (GLT)-based wood adhesives. **Figure 4** shows the synthesis of hyperbranched poly(amine-ester).

Previous studies by Lei [21] indicated the use of GL/PF/pMDI formulation with calcium lignosulfonate, and it turned out that the molecular mass decreased significantly during thermal treatment at 170°C for 90 min and pH of 12.7. This has been done to reduce the molecular mass of the lignin. Some linkages of lignin are broken by thermal treatment, and positions that are more reactive will be obtained. This indicates that the thermal treatment will make the material more suitable for such reactions. In this work, the two types of low molecular mass lignins extracted from oil palm empty fruit bunch were used. These are (i) kraft lignin and (ii) organosolv lignin. The adhesive resin properties are shown in **Table 1**.

The quality of wood and how the resin penetrates through the wood surface are important factors that affect the assessment of the adhesive joint in plywood. The bonding strength formed by TGKL and MTGK adhesives was studied via tensile strength test on wood substrates. For a comparative study, the same test was implemented for synthetic commercial phenol-formaldehyde resin. The data is displayed in **Table 2**.

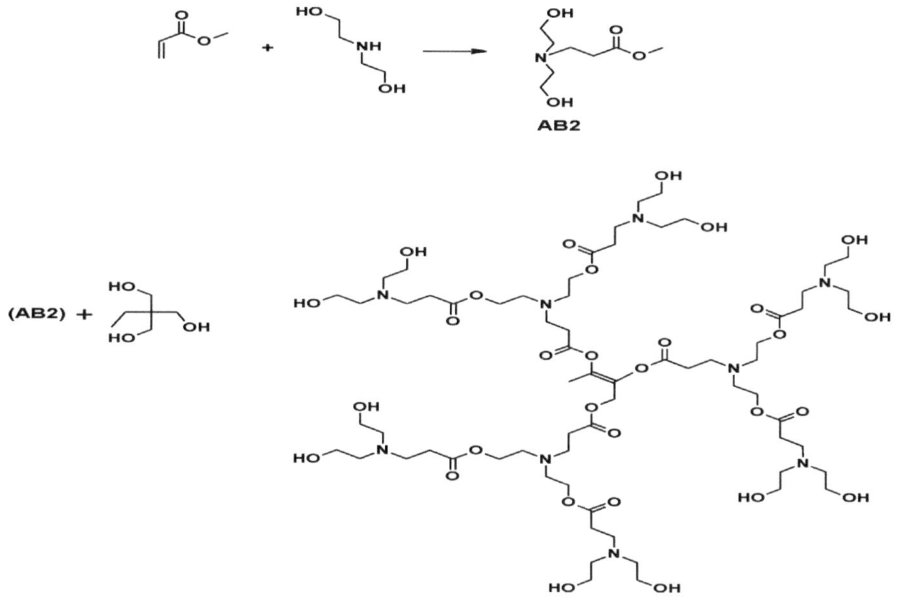

Figure 4.
Synthesis of hyperbranched poly(amine-ester).

Resin type	Viscosity at 30°C, 100 rpm (cP)	Solid content[a] (%)	pH	Gel time at 100°C (s)
CPF	190	59.80	11.30	480
TGKL	65	39.26	9.46	164
MTGKL	108	53.08	9.30	235

[a]*Solid content of the resins was determined by measuring the weight before and after removing the solvent by heating at 105°C to constant weight.*
%Solid content = (weight of the solid resin/weight of the solution) × 100.

Table 1.
Variation of the physical properties of CPF, TGKL, and MTGKL resins.

The maximum load, tensile strength, elastic modulus, and elongation at break of panels are direct measures of the performances. From **Table 2**, the trend shows that the dry MTGKL resin has the tensile strength of 39.72 MPa and elongation at break of 21.91%. This result indicates that MTGKL resin is stronger than that of the CPF resin. TGKL has the tensile strength of 28.79 MPa and elongation at break of 20.61%. That is also stronger than that of the CPF resin but lesser than MTGKL resin. The results of MTGKL and TGKL resins are within the requirement of the relevant international standard as per British Standard for dry tensile strength of plywood (≥0.35 MPa) [40].

Interestingly when comparing TGKL with MTGKL resins, the water resistance of MTGKL resin is improved after modification with poly(amine-ester). This is evident in the results of tensile strength (2.72 MPa) after soaking the plywood samples of MTGKL resin in tap water for 24 h at room temperature, where no delamination occurred in any of the specimens. However, delamination was observed in the plywood samples of TGKL resin after soaking in water for a period not exceeding 1 h, indicating the lack of water resistance for TGKL resin when it was used to bond plywood. This suggests that the modification of tannin by adding oligomeric precursors (hydroxyl-terminated) of a hyperbranched poly(amine-ester) is successful to boost the network structure of natural phenolic resin to prepare a water-resistant resin.

Resin type		Tensile strength, MPa		Elastic module, MPa	Elongation at break (%)
	DRY	Cold water soaking, 24 h	Boiling water soaking, 2 h		
CPF	1.39	---	---	56.61	12.45
	---	1.25	---	41.10	15.84
	---	---	1.01	26.78	17.29
TGKL	2.89	Delamination	---	125.27	20.61
MTGKL	3.97	---	Delamination	185.30	21.91
	---	2.72	---	143.66	18.20

Table 2.
Tensile strength, elastic modulus, and elongation at break of plywood using CPF, TGKL, and MTGKL resins.

The improvement of water resistance of MTGKL resin is assumed to be due to the reaction of the terminal units of glutaraldehyde with the hydroxyl groups of tannin and with $-NH_2$ and $-COOH$. Other exposed groups lead to an increase in the cross-linking density within the resin in the hot pressing process, and consequently, there will be an increase in the mechanical properties. In addition, the presence of furfuryl alcohol as a cross-linking agent and its participation in increasing the cross-linking reaction help to improve the water resistance of MTGKL-based wood adhesives. The whole process is represented in **Figure 5**.

4.2 A combination of lignin polyol-tannin adhesives and polyethylenimine (PEI) as fillers

Protein adhesive, which was secreted via marine mussels, called marine adhesive protein (MAP), is a good example of a renewable resource and free formaldehyde adhesive [42]. MAP has the ability to form strong linkages on wet surfaces and thus could be used as a strong and water-resistant wood adhesive [42]. However, they are costly and not readily available. MAP contains two functional groups (catechol and amino groups). The different reactions between the catechol group and the amino group lead to the cross-link and solidification of the MAP, thereby converting the MAP to a very strong and very water-resistant adhesive [42].

Condensed tannins and lignin are one of the few natural polymers containing a catechol moiety [43]. Previous studies showed the possibility of using a combination of condensed tannins or lignin and polyethylenimine to synthesize wood adhesives. **Figure 6** shows representative structures of polyethylenimine. This adhesive has high shear strength and exhibits significant water resistance [43]. In this work, the wood adhesive system, which consists of a tannin-glyoxalated lignin polyols (TGLP) and PEI, was evaluated for plywood and compared to plywood produced with a glyoxalated lignin/tannin and conventional phenol-formaldehyde resin.

It was observed that an increase in the total solid content of the TGLP-PEI adhesives leads to an increase in its viscosity. Moreover, when the total solid contents in adhesives reached 55%, the TGLP-PEI adhesives became too viscous. It can clearly be seen from **Figure 7** that the total solid contents increased from 48.54 to 56.92% when the ratio of PEI was changed from 10 to 20% in TGLP resin. The tensile strengths (dry, WSAD, and BWT/dry) of wood composites bonded with the adhesives have revealed a dramatic increase in its values (63.04, 59.48, and 53.53 MPa), respectively, and a linear relationship among all values. The TGLP-PEI

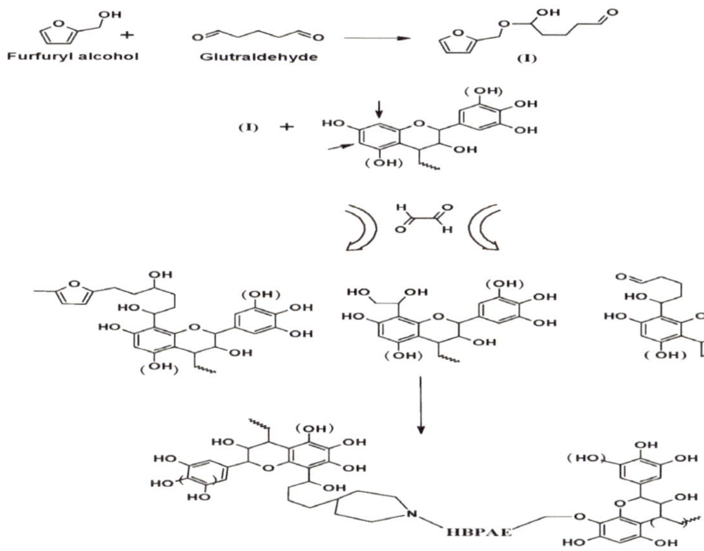

Figure 5.
Proposed mechanism for the participation of formulation components in cross-linking reactions and improvement of water-resistant properties for the resulting modified tannin-glyoxalated lignin [41].

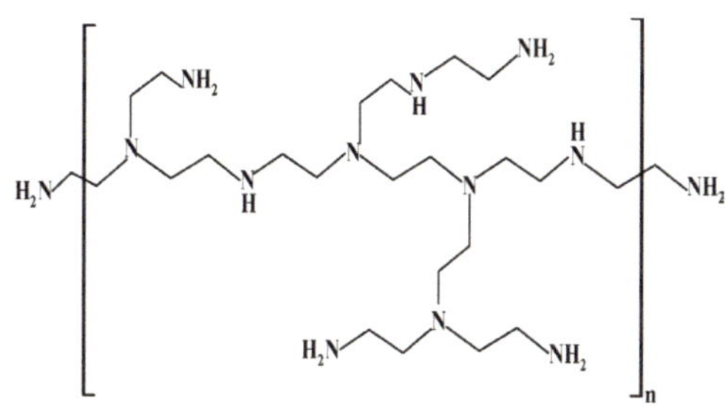

Figure 6.
Representative structures of polyethylenimine (PEI).

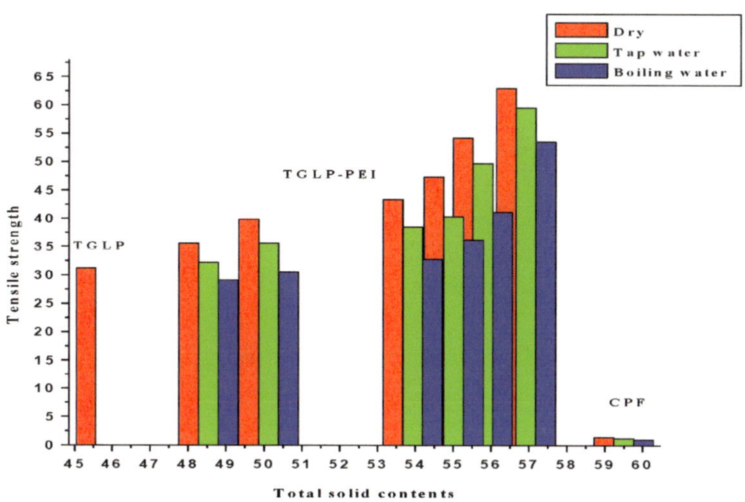

Figure 7.
Effect of total solid content of TGLP-PEI adhesives on tensile strength (dry, tap water soaking for 24 h, and boiling water soaking for 2 h) in comparison with TGLP and CPF.

adhesives with total solid contents (56.92%) were applied to plywood and used in subsequent experiments.

The production of wood composites does not favor the use of wood adhesives with low solid contents due to higher energy consumption and comparatively longer time of water evaporation through hot press. In the current study, a significant amount of the total solid contents of adhesives (56.92%) was obtained during the modification of lignin polyol-tannin resins with polyethylenimine. There was a substantial difference in the solid contents of modified (56.92%), unmodified adhesives (45.82%), and the commercial phenol-formaldehyde (59.50%) under the same experimental conditions. The solid content of the TGLP resin (45.82%) was lower than the solid contents of the CPF resin (59.50%). However, the solid contents of the TGLP-PEI resins were almost comparable to the CPF resin, especially when the ratio of PEI was 20% in TGLP-PEI resin. It is known that CPF resins contain large amount of urea, and for this reason, the solid content value is very high [44].

To date, the reactions between lignin and tannin and PEI are not fully understood. Some of the proposed potential reactions are given in **Figure 8**. The nature of TGLP-PEI resin curing mechanisms is comparable to the quinone-tanning methods as reported elsewhere. Moreover, the reaction mechanisms between lignin and PEI are mostly similar to the possible reactions between tannin and PEI. At elevated temperatures, the catechol moiety (2) in the demethylated lignin is highly vulnerable to oxidation process, leading to the formation of quinones (3). This occurs at higher temperatures (140°C) using a hot press during the formation of wood composites. The quinones (3) possibly react with the amino groups present in PEI structure, thus forming Schiff bases (4 and 5). Some other possible reactions between quinones and amino groups of PEI (such as Michael addition reaction) may also occur to form (6) further Schiff bases (7 and 8). A similar kind of reaction occurs during the oxidation process to form phenolic hydroxide in lignin structure during hot press, which further results in the formation of quinone. It further supports the increase of the reactions with amino groups of PEI to form Schiff bases.

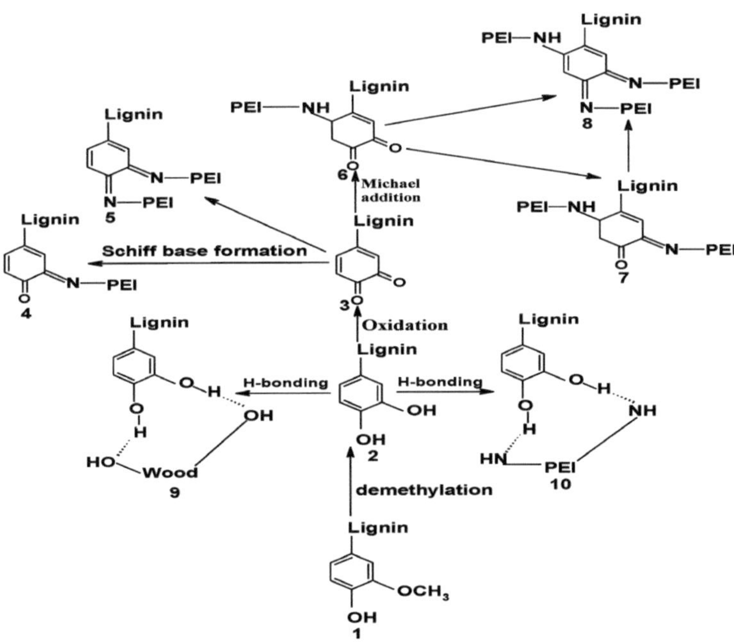

Figure 8.
Some possible reactions between tannin and PEI.

It is well known that wood has lignin, which contains phenolic hydroxyl groups, and these groups can be oxidized to quinone during the hot press; hence they cannot be excluded from the covalent bond formation between TGLP-PEI adhesive and wood. In addition to the reaction of the quinone, the catechol moiety can react with the amino groups of PEI and form strong hydrogen bonds. Furthermore, it can also form strong hydrogen bonds with the hydroxyl groups in the wood components (see a representative structure 9 and 10 in **Figure 8**). It can be inferred from these reactions that the curing reactions through the hot press process lead to the formation of highly cross-linked TGLP-PEI network polymers, as well as water-resistant adhesives. These results are in close agreement to mechanical tests and proven through soaking plywood samples bonded with TGLP-PEI adhesives in the tap water and boiling water.

4.3 Influence of bentonite (BT) clay on lignin

Improving the polymer properties through incorporating a second component such as fillers to obtain new materials has become common. Inorganic additives are frequently used in the wood adhesive industry that improves the mechanical and thermal properties of resins [36]. Recently, the utilization of the montmorillonite to modified phenolic resin attracts significant interest of researchers to improve some properties such as stiffness and toughness [45]. In this work, wood adhesive system, consisting of a tannin-glyoxalated lignin polyol, (TGLP) was modified using bentonite and organo-bentonite clay and then evaluated on plywood.

Scanning electron microscope (SEM) is a significant technique used to interpret changes in morphology of bentonite during a modification process by surfactants. SEM micrographs of bentonite before and after modification with octadecylamine (ODA) as well as the SEM images of composites with adhesive are shown in **Figure 9**. The corresponding elemental analysis data are summarized in **Table 3**. It is worth mentioning that there are no substantial differences in morphology between natural and organic bentonite.

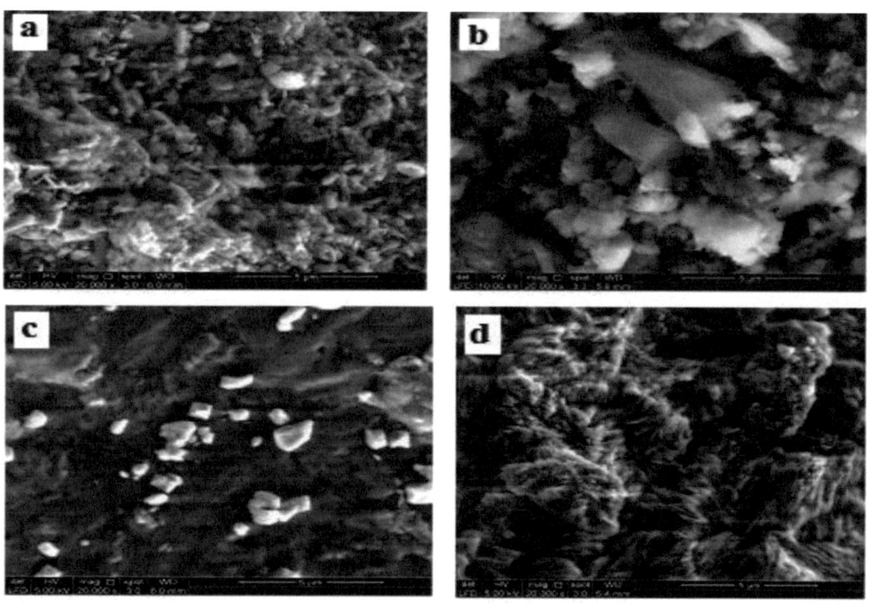

Figure 9.
SEM micrographs of (a) unmodified BT, (b) ODA-modified BT, (c) TGLP-BT, and (d) TGLP-ODA-BT composites.

Sample	C (w%)	O (w%)	Si (w%)	Al (w%)
BT	12.79	61.05	17.96	5.73
ODA-BT	21.12	68.84	6.22	1.97
TGLP-BT	22.42	67.14	3.95	1.21
TGLP-ODA-BT	23.22	66.39	0.72	0.64

Table 3.
EDX elemental analysis of unmodified BT, ODA-BT, and composites.

The unmodified bentonite (**Figure 9a**) demonstrated huge aggregated morphology with large-sized particles in the structure. Some fragments of small-sized particles in BT structure were relatively regular. This may be due to the interactions of face-to-face and face-to-edge of particles. Comparing with the unmodified BT, the BT-ODA (**Figure 9b**) showed smaller particles and irregular shapes as well as more severely aggregated particles than BT. In addition, the SEM images also revealed the expansion that took place in layers of silica and a more open structure in the ODA-BT. As a result, further exfoliation during composite fabrication is expected.

SEM analysis of BT-TGLP composites in **Figure 9c** exhibited the lack of compatibility between the unmodified bentonite and matrix resin, which could lead to the formation of tactoids. During the modification of organic bentonite using appropriate modifiers, the interaction between the matrix resin and the organo-bentonite would be sufficient for a certain degree of exfoliation and/or intercalation even when forming the tactoid. SEM micrograph of ODA-BT containing the composite resin is given in **Figure 5d**. As shown in **Figure 9d**, ODA-BT particles were dispersed in the TGLP matrix, and the resulting composite structure (TGLP-ODA-BT) demonstrated a highly homogenous distribution.

The elemental analysis by EDX of the BT, ODA-BT, and composites is shown in **Table 3**. The energy spectra indicate that the elements present are carbon (C), silicon (Si), aluminum (Al), and oxygen (O). From **Table 3** the BT sample has the highest content of silicon (Si) and aluminum (Al). It is clear that the bentonite is a class of aluminum silicate minerals. For ODA-BT, it was observed that the amount of carbon (C) element (21.12 w%) has significantly increased compared with BT sample (12.79 w%), indicating the presence of alkylammonium cations as intercalant group in the bentonite gallery that additionally supports the modification that occurred. It is noteworthy to point out that the percentage of all the other elements such as Si and Al decreased significantly and this can be due to the presence of carbon (C).

The average tensile strength of unfilled TGLP resin and composite with corresponding bentonite and organo-bentonite loadings was determined from the tests conducted on at least five plywood specimens under dry conditions as shown in **Figure 6**. The tensile strength of TGLP composite has been considerably improved by incorporating both BT and ODA-BT.

The tensile strength of TGLP composite increases nonlinearly with BT and ODA-BT content. From **Figure 10**, the dry tensile strength was more affected by ODA-BT than BT. The reason in the strengthening of tensile properties is probably due to intercalation/exfoliation of the bentonite structure in the phenolic matrix of adhesives. As the silicate layer in bentonite structure has excellent mechanical characteristics compared with TGLP, the enhanced mechanical properties of resin may be linked to the dispersion level of the silicate layers into the matrix of resin. It is believed that the dispersion process of clay in polymer chain provides a large interfacial interaction that may cause a restriction on the mobility.

There are multiple factors that affect the mechanical properties including the ratio of filler, dispersion of the filler, and the adhesion at the filler matrix interface

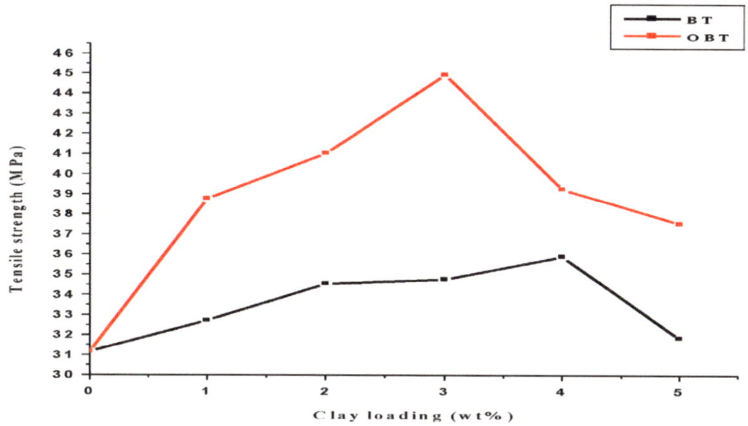

Figure 10.
Effect of the clay loading of TGLP adhesives on tensile strength.

[46]. In addition, the exfoliation degree of the silicate layer in the polymer affects the modulus of composites. This reveals that the significant increase of the tensile strength modulus through the incorporation of low contents of ODA-BT (1, 2, and 3 wt%) can be due to the uniform dispersion of nanoparticles at such a low content. High content of particles (4 and 5 wt%) reduced the dispersion and thus restricts the improvement of tensile strength (**Figure 10**). The elastic modulus has the same sensitivity toward dispersion. Dispersion of filler particles has less effect when BT is used; the limitations of tensile strengths show at a loading of 5% (**Figure 10**).

The decrease of the tensile strength at high content of BT and ODA-BT can be attributed to the agglomeration of particles. An agglomeration of particles could also serve as crack initiation sites, as well as the nonuniform distribution. Through the results, it could be concluded that the use of ODA-BT is better than BT to improve the mechanical properties, particularly tensile strength when used in low concentrations. This can possibly be due to lower silicate layers in the ODA-BT than BT because of the organic modification of bentonite that has a positive effect on ODA-BT, thus increasing the cross-linking density.

With the modification of bentonite, the ODA-BT became more compatible with the TGLP resin chains, due to the decrease of surface energy of bentonite layers; thus the surface polarity of ODA-BT is compatible with the resin polarity. The ODA-BT with lowered surface energy has more ability to interact and intercalate within the interlayer space of resin than BT and thus improves the various properties of the resin.

5. Conclusions

Composite materials have a pivotal role in industries that are now considered the most progressive worldwide. Polymer composites are a mixture of polymers with inorganic or organic fillers that have certain geometric forms. Wood adhesives now represent a vital aspect of the wood-based panel industry, and synthetic condensation resins have become widely used. At present, synthetic adhesives based on formaldehyde such as PF, UF, and MF are predominantly used for wood composite production. A variety of methods have been used to improve the performance of wood adhesives. These methods are widely used in the industry; they include the simple addition of fillers. In this chapter, it was demonstrated that a combination of lignin polyol-tannin (TGLP) and polyethylenimine (PEI) was an excellent alternative for marine adhesive protein (MAP). The results revealed that the increase in PEI ratio led to an increase of the solid content of TGLP-PEI adhesives. The TGLP adhesives had poor water resistance compared with TGLP-PEI adhesive despite the high tensile strength (31.1 MPa). This is clearly shown when soaking the plywood specimens in the tap water and boiling water, where delamination occurred. However, at 10% of PEI with TGLP resin, the delamination of plywood specimens did not occur.

There are not a lot of studies conducted on the influence of clays on wood adhesives prepared from biomass. The mechanical and thermal characteristics of TGLP system were studied. The modification of bentonite clay was implemented using octadecylamine (ODA) salt. The mechanical properties (tensile strengths) of TGLP composites have been considerably improved by incorporating both BT and ODA-BT. The tensile strengths revealed that the use of low contents of clay is better than using high clay content that could be due to the decrease in dispersion at high contents of clay. These results are particularly clear when using modified bentonite with ODA. This is probably due to the decrease of some silicate layers in the ODA-BT compared with BT.

Acknowledgements

The author is grateful for the financial support of this chapter from the University Science Malaysia through the USM Research University Grant 1001/PKIMIA/854002 and to the laboratories of Material Research Directorate in the Ministry of Higher Education and Scientific Research, Baghdad, Iraq.

Author details

Abbas Hasan Faris
Materials Research Directorate, Ministry of Higher Education and Scientific Research, Baghdad, Iraq

*Address all correspondence to: abbas_hf@yahoo.com

References

[1] Armentano I, Dottori M, Fortunati E, Mattioli S, Kenny JM. Biodegradable polymer matrix nanocomposites for tissue engineering: A review. Polymer Degradation and Stability. 2010;**95**:2126-2146

[2] Cosoli P, Scocchi G, Pricl S, Fermaglia M. Many-scale molecular simulation for ABS-MMT nanocomposites: Upgrading of industrial scraps. Microporous and Mesoporous Materials. 2008;**107**:169-179

[3] Ma H, Xu Z, Tong L, Gu A, Fang Z. Studies of ABS-graft-maleic anhydride/clay nanocomposites: Morphologies, thermal stability and flammability properties. Polymer Degradation and Stability. 2006;**91**:2951-2959

[4] Pandey JK, Reddy KR, Kumar AP, Singh RP. An overview on the degradability of polymer nanocomposites. Polymer Degradation and Stability. 2005;**88**:234-250

[5] Transparency MR. Phenolic Resins (Resol, Novolac and Others) Market for Wood-Adhesives, Molding Compounds, Laminates, Insulation and Other Applications: Global Industry Analysis, Size, Share, Growth, Trends and Forecast. 2013

[6] Baekeland LH. The synthesis, constitution, and uses of bakelite. Journal of Industrial and Engineering Chemistry. 1909;**1**:149-161

[7] Giannelis EP. Polymer-layered silicate nanocomposites: Synthesis, properties and applications. Applied Organometallic Chemistry. 1998;**12**:675-680

[8] Pavlidou S, Papaspyrides CD. A review on polymer-layered silicate nanocomposites. Progress in Polymer Science. 2008;**32**:1119-1198

[9] Thostenson ET, Li C, Chou TW. Nanocomposites in context. Composites Science and Technology. 2005;**65**:491-516

[10] Alexandre M, Dubois P. Polymer-layered silicate nanocomposites: Preparation, properties and uses of a new class of materials. Materials Science and Engineering. 2000;**28**:1-63

[11] Murray HH. Traditional and new applications for kaolin, smectite, and palygorskite: A general overview. Applied Clay Science. 2000;**17**:207-221

[12] Lebaron PC, Wang Z, Pinnavaia TJ. Polymer-layered silicate nanocomposites: An overview. Applied Clay Science. 1999;**15**:11-29

[13] Sansuri AJB. Super absorbent polymer composites [thesis]. Pahang: University Malaysia; 2010

[14] Giannelis EP. Polymer layered silicate nanocomposites. Advanced Materials. 1996;**8**:29-35

[15] Singla P, Mehta R, Upadhyay SN. Clay modification by the use of organic cations. Green and Sustainable Chemistry. 2012;**2**:21-25

[16] Van Olphen H. An Introduction to Clay Colloidal Chemistry. New York: Wiley; 1977

[17] Maiti P, Yamada K, Okamoto M, Ueda K, Okamoto K. New polylactide/layered silicate nanocomposites: Role of organoclays. Chemistry of Materials. 2002;**14**:4654-4661

[18] Ahmad MB, Hoidy WH, Ibrahim NAB, Al-Mulla EAJ. Modification of montmorillonite by new surfactants. Journal of Engineering and Applied Science. 2009;**4**:184-188

[19] Chigwada G, Wang D, Jiang DD, Wilkie CA. Styrenic nanocomposites

prepared using a novel biphenyl-containing modified clay. Polymer Degradation and Stability. 2006;**91**:755-762

[20] Gu JY. Adhesive and Paint. Beijing: China Forestry Publishing House; 1998

[21] Lei H. Synthetic and natural materials for wood adhesive resins [thesis]. Epinal: University Henri Poincare-Nancy; 2009

[22] Ma WW. The preparation of urea-formaldehyde resin modified by oxygenated amylum. Chemistry and Adhesion. 1995;**3**:147-148

[23] Shi YM. Development of the low cost urea-formaldehyde resin adhesive. China Adhesive. 2003;**12**:30-32

[24] Mosiewicki M, Aranguren MI, Borrajo J. Thermal and mechanical properties of wood flour/tannin adhesive composites. Journal of Applied Polymer Science. 2004;**91**:3074-3082

[25] Kargarfard A, Jahan-Latibari A. Application of recycled polyethylene in combination with urea-formaldehyde resin to produce water resistant particleboard. In: Proceedings of the 55th International Convention of Society of Wood Science and Technology. Beijing, China; 2012. pp. 1-7

[26] Du GB, Wu LZ, Yang QX. Development on mineral filler for UF resin. Adhesion. 1995;**16**:9-12

[27] Cavicchi RE, Silsbee RH. Coulomb suppression of tunneling rate from small metal particles. Physical Review Letters. 1984;**52**:1453

[28] Lin QJ, Yang GD, Liu JH. Study on the property of nano-SiO_2/urea-formaldehyde resin. Scientia Silvae Sinica. 2005;**41**:129-135

[29] Yang GD, Lin QJ, Liu JH. The effect of nanometer silicon dioxide on properties of UF resin. Journal of Fujian Forestry College. 2004;**24**:114-117

[30] Yang G, Lin QJ, Liu JH. Study on using nanometer SiO_2 for modifying UF resin. China Wood Industry. 2004;**18**:7-9

[31] Yu HW, Fu SY, Wen GF. The effect of nanometer calcium carbonate on properties of UF resin. China Adhesives. 2002;**11**:22-24

[32] Zhang W, Ma Y, Xu Y, Wang C, Chu F. Lignocellulosic ethanol residue-based lignin–phenol–formaldehyde resin adhesive. International Journal of Adhesion and Adhesives. 2013;**40**:11-18

[33] Lv JK, Ke YC, Qi ZN. Synthesis and mechanical properties of epoxy/clay nanocomposites. Acta Materials Composed Sinica. 2002;**19**:117-121

[34] Tien Y, Chen TK, Wei KH. Synthesis and characterization of novel segmented polyurethane/clay nanocomposites. Polymer (Guildf). 2000;**41**:1345-1353

[35] Choi MH, Chung IJ, Lee JD. Morphology and curing behaviors of phenolic resin-layered silicate nanocomposites prepared by melt intercalation. Chemistry of Materials. 2000;**12**:2977-2983

[36] Pizzi A, Mittal KL. Handbook of Adhesive Technology. 2nd ed. New York: Marcel Dekker; 2003

[37] Magnusson H, Malmström E, Hult A. Structure buildup in hyperbranched polymers from 2,2-bis(hydroxymethyl) propionic acid. Macromolecules. 2000;**33**:3099-3104

[38] Fang K, Xu Z, Jiang X, Zhang X, Fu S. Preparation and characterization of hyperbranched polyesteramides. Polymer Bulletin. 2008;**60**:533-543

[39] Khan JA, Kainthan RK, Li MG. Water soluble nanoparticles from

PEG based cationic hyperbranched polymer and RNA that protect RNA from enzymatic degardation. Biomacromolecules. 2006;**7**:1386-1388

[40] British Standard. Specification for Bond Performance of Veneer Plywood. London 9, 1985; part 8

[41] Li X, Essawy H, Pizzi, Delmotte L, Rode K, Le Nouen D, et al. Modification of tannin based rigid foams using oligomers of a hyperbranched poly(amine-ester). Journal of Polymer Research. 2012;**19**:21

[42] Waite JH. The phylogeny and chemical diversity of quinone-tannin glues and varnishes. Comparative Biochemistry and Physiology. 1990;**B 97**:19-29

[43] Li K, Geng X, Simonsen J, Karchesy J. Novel wood adhesives from condensed tannins and polyethylenimine. International Journal of Adhesion and Adhesives. 2004;**24**:327-333

[44] Zhao Y. Development of bio-based phenol formaldehyde resol resins using mountain pine beetle infested lodgepole pine barks [thesis]. University of Toronto; 2013

[45] Jiang W, Chen SH, Chen Y. Nanocomposites from phenolic resin and various organo-modified montmorillonites: Preparation and thermal stability. Journal of Applied Polymer Science. 2006;**102**:5336-5343

[46] Wang K, Wu J, Chen L, He C, Toh M. Mechanical properties and fracture behavior of epoxy nanocomposites with highly exfoliated pristine clay. In: ANTEC Conference Proceedings. Chicago; 2004. pp. 1820-1824

Chapter 2

Wood Adhesive Fillers Used during the Manufacture of Wood Panel Products

Long Cao, Xiaojian Zhou and Guanben Du

Abstract

During the manufacture of wood panel products, fillers are commonly added to wood adhesives to lower costs and give body to liquid adhesives and also reduce undesired flow or overpenetration into wood. The fillers used in wood adhesives are often neutral or weakly alkaline compounds that typically require no chemical reaction with curing agent, or other components. Fillers are mixed with other components prior to the application of resin on the surface of wood, wood veneer, or wood flakes. Fillers can be either organic (e.g., rye, wheat, walnut shell, and wood flours), or inorganic (e.g., calcium carbonate, calcium sulfate, aluminum oxide, or bentonites). Overall, fillers are low-cost materials for improving the properties of wood or even give it new functions.

Keywords: fillers, wood adhesives, performances, advantages, applications

1. Introduction

Fillers are solid additives that are primarily used to lower the cost and give body to liquid adhesives or reduce undesired flow or overpenetration into wood. This leads to an improvement in properties of the adhesives and gives rise to new functions [1]. Fillers usually increase the rigidity of cured adhesives. They may also modify the coefficient of thermal expansion of a film to approximately that of the adjacent adherends. This can reduce thermal stresses in the joint generated during cooling following heat-curing conditions or when thermally cycled during service. Fillers, normally, are neutral or weakly alkaline compounds and do not chemically react with adhesives, curing agents, or other components in wood adhesive system.

2. Filler species

There are many kinds of adhesives for manufacturing of wood panel products, and the type and the amount of filler greatly affect its performance [2]. The choice of filler depends on the materials and application, making fillers one of the most important components of adhesives during the production of wood-based boards. Adding an appropriate filler to an adhesive can reduce the amount and cost of glue and also improve the performance of the adhesive. Commonly used adhesive fillers are organic, such as flour, soybean powder, wood powder, and bark powder or other agroindustrial wastes (**Table 1**), such as palm kernel and starch material, etc. In addition, inorganic materials such as metal powders, metal oxides, and minerals, have also been used as adhesive fillers, typically to improve compression strength and dimensional stability.

Types	Amounts (%)	Function	Main application in wood adhesive	Application fields
Flour, wood powder, soybean powder, bark powder	10–30	Decrease cost, increase ductile behavior, avoid overpenetration and heterogeneous spread	Urea-formaldehyde, phenol-formaldehyde, melamine-formaldehyde resin, polyvinyl acetate resin	Wood-based panels, wood-glued products
Copper oxide, magnesium oxide, magnesium hydrate, iron sesquioxide, titanium dioxide, chromium hemitrioxide, zinc oxide, copper powder, silver powder, magnesium carbonate, iron powder	25–100	Improve compression, hardness, thermal conductivity, electrical conductivity, and processability	Epoxy, rubber, polyurethane glue, phenol-formaldehyde resin, polyvinyl acetate resin	Wood structure products
Calcium carbonate, cement, clay, talcum powder	25–100	Reduce cost and curing shrinkage and increase compression and hardness	Polyurethane, epoxy, urea-formaldehyde, phenol-formaldehyde, polyvinyl acetate resin	Wood structure products
Carborundum, mica powder	<100	Reduce cost and increase physical properties	Epoxy, phenol-formaldehyde resin	Wood structure products
Synthetic cement, Ganister sand	10–20	Improve the fluid flow performance	Polyurethane, epoxy resin	Wood structure products
White carbon black	20–100	Improve the mechanical properties	Rubber, epoxy resin	Musical device
Land plaster	10–100	Reduce cost and increase viscosity	Epoxy resin	Wood bonding and coating
Kaolin	<10	Improve thixotropy and fluid flow performance	Epoxy resin, urea-formaldehyde resin, polyvinyl acetate resin	Wood-based panels
Graphite	<50	Increase thermal performance, lubricity, and wear resistance	Epoxy, rubber resin, polyvinyl acetate resin	Musical device
Asbestos powder, glass fiber, talcum powder	20–50	Improve impact strength and heat resistance	Rubber, epoxy, urea-formaldehyde resin, polyvinyl acetate resin	Musical device and coating

Table 1.
The types, amount, and function of fillers in wood adhesives.

According to their color, fillers can be divided into white fillers and color fillers. They can also be divided according to their preparation method into natural fillers and synthetic fillers. They can be divided according to their function into temperature-resistant fillers, conductive fillers, and anti-sink fillers. According to their particle sizes, they can be divided into natural fillers, ultra-fine fillers, and nanoscale fillers. According to their composition, they can be divided into compound fillers and mixture fillers.

3. The advantages of fillers for wood adhesives

The adhesive composition mainly includes a matrix material, curing agent, toughening agent, diluent, filler, and modifier. However, fillers are solid materials that do not chemically react with the adhesive component but can change its performance [3, 4]. The main functions are summarized in **Table 1** and further discussed in this section.

3.1 Increase the mechanical properties of the adhesive

Many polymers have weak intermolecular interactions and low cohesive energies, so their mechanical properties are inferior to other materials. Fillers with an appropriate particle size can enhance the adhesive strength, and the active surface of filler particles can be used to cross-link several large molecular chains to form a network structure. When one molecular chain is stressed, the stress can be dispersed and transferred to other molecules through cross-linking. Even if one chain fractures, the other chains remain intact, and it is unlikely that the entire structure will immediately fracture, leading to a substantial improvement in the mechanical properties of adhesives. Commonly used fillers, such as flour, metal powders, and metal oxides, can improve the compression strength of adhesives and their dimensional stability. Adding carbon black, silica, or calcium carbonate into silicone and rubber glue can improve the tensile strength, hardness, and wear resistance, etc.

3.2 Give new functions of adhesives

Conductive and magnetic adhesives are obtained by the addition of silver powder and carbon-based iron powder into adhesives, respectively. The thermal conductivity of adhesives can be improved using copper powder, aluminum powder, alumina, and magnesium oxide as fillers. The thermal conductivity adhesives can be widely used in microelectronic assembly and bonding electronic products instead of spot welding. The magnetic adhesives can be improved production efficiency because of simple operation process in electrical machinery bonding field. In epoxy resins, zinc chromate and $Zr(SiO_3)_2$ can help retain strength and reduce water absorption. Flame retardant powders such as aluminum hydroxide can improve the flame retardancy of an adhesive. Some fillers can also improve the resistance of adhesive joints to moisture and heat aging and salt spray.

3.3 Reduce joint stress

Fillers can prevent local overheating near the bonding interface because the curing reaction is exothermic. In most cases, the curing shrinkage of wood adhesives often occurs during the glue bonding process, but filler can be used to adjust the shrinkage rate. The addition of wheat flour can reduce cracking, which is caused by curing shrinkage of urea-formaldehyde resins. The proper selection of filler can reduce the difference between the thermal expansion coefficient and the expansion rate between adhesives and bonded materials. Additionally, it can reduce the internal stress of joints, the thermal expansion coefficient, and curing shrinkage ratio of adhesives.

Colloids form during the curing process due to chemical reactions and cause volume shrinkage. Thermal shrinkage will also occur due to the different thermal expansion coefficient of the adhesive. These two types of shrinkage will produce internal stresses in the rubber layer, resulting in stress concentration, cracking, or joint damage of the rubber layer, which directly affects the service life of rubber joints. Filler can be used to adjust shrinkage during curing, reduce the difference of thermal expansion coefficients between the wood adhesive and the object being glued, and can also prevent cracks from extending. Thus, fillers can significantly improve the bonding strength, especially the shear strength at high temperatures.

3.4 Improve operation process

Fillers in adhesives can adjust the curing speed, prolong the pot life, and facilitate manufacturing. During plywood manufacturing, wheat flour added into urea-formaldehyde resin can increase the viscosity of an adhesive to prevent it from excessively penetrating into wood pores. Fillers can also improve the thixotropy of liquid glues to control their fluidity, adjust the curing speed, extend the service life, and facilitate operation and construction.

Normally, adhesive bonding strength, adhesion, and heat resistance significantly increase when a certain amount of filler is added, especially polar fillers, such as metal powders, metal oxides, and minerals. This reduces adhesive curing shrinkage and coefficients of thermal expansion.

The addition of asbestos wool and glass fiber has been shown to improve the impact strength. Quartz, porcelain, and iron powders can increase the hardness and compression resistance of adhesives, while graphite and talcum powders can improve wear resistance. Alumina and titanium dioxide can increase the bonding strength. Flour is the most widely used filler in wood adhesives in the wood panel industry and is used to improve the mechanical properties, shrinkage, expansion behaviors, and other physical performances at the glue-wood interface. Most importantly, the use of flour reduces the cost.

An appropriate proportion of filler should be used to provide the desired function and to ensure the overall superior performance of an adhesive. A high proportion of filler will increase the viscosity of the adhesive, making it difficult to control and stir, leading to inferior wettability and a low bonding strength. It may also reduce the de-lamination strength and increase wood failure. Overall, the purpose of fillers is to enhance the physical and mechanical properties of wood-based panels.

Filler selection should meet the following requirements:

1. Non-toxic
2. Unreactive toward other components in the wood adhesive system
3. In a specific physical state, such as uniform particle size
4. Low-cost, a wide range of sources, and convenient processing
5. Should not contain moisture, grease, or harmful gases; moisture absorption is not easily changed

DOI: http://dx.doi.org/10.5772/intechopen.91280

Figure 1.
The function and selection of fillers.

6. Easily dispersed and has good lubrication with adhesive

7. Should meet the specific requirements of the adhesive, such as electrical conductivity, heat resistance, etc.

The selection of appropriate fillers in wood adhesive systems is important because different fillers have different effects, as show in **Table 1** and **Figure 1**.

4. Applications of fillers

There are many different types of fillers, which have different applications. The most widely used fillers include natural calcium carbonate, barite powder, quartz powder, talc powder, kaolin, mica powder, attapulgite (aluminum-silicon-magnesite containing water), and flours of different renewable biomaterials.

Nanomaterials have quantum size effects, small size effects, surface effects, and macroscopic quantum effects. The addition of small amounts of nanoscale powders into an adhesive can significantly increase its viscosity, improve the bonding strength, and prevent caking. For example, it was found that nanoscale calcium carbonate prolonged the curing time of resin and the content of free formaldehyde decreased as the amount of calcium carbonate increased. Nanoscale montmorillonite was also shown to increase mechanical properties and reduce formaldehyde emission [5–7].

4.1 Calcium carbonate

Calcium carbonate ($CaCO_3$) is an odorless white powder and is one of the most widely used fillers [8]. Lightweight precipitated calcium carbonate can be synthesized by a chemical method with a whiteness of 90% and a relative density of 2.6 g/cm^3. Heavy calcium carbonate is composed of natural calcite, limestone, chalk, and shells, which are ground to certain fineness by a mechanical method.

In the adhesives industry, calcium carbonate is widely used as a filler because of its low price, non-toxicity, white color, abundant resources, easy mixing in formulas, and stable performance. The addition of nanoscale calcium carbonate to an adhesive can enhance the mechanical strength and increase the transparency, thixotropy, and spreading smoothness. Additionally, the adhesive easily provides a shielding effect, leading to an anti-UV aging effect, as well as an improvement in its mechanical strength.

4.2 Kaolin

Kaolin, usually called clay or china clay, has a main mineral component of kaolinite, which is a variety of crystal rock with the molecular formula

$Al_2O_3 \cdot SiO_2 \cdot nH_2O$. Kaolinite has a flake structure and can be divided into calcined kaolin and washed kaolin. Calcined kaolin generally has a higher oil absorption, opacity, porosity, hardness, and whiteness than washed kaolin.

Kaolin normally forms an unstable structure in water because of its charge distribution, with positively charged sheet edges and a negatively charged surface. If the kaolin dosage is high, it will form a gel, preventing an adhesive from flowing [9]. Clay is sometimes added to epoxy resins to thicken or modify coefficients of thermal expansion.

4.3 Renewable bio-based materials

Flour and other renewable bio-based materials include wood powder, starch, protein, and lignocellulose as well as the agroindustrial wastes. Adding a small amount of starch into wood adhesives can significantly increase the viscosity and effectively improve the solid content and initial viscosity of the adhesive [10–12]. Oxidized starch and palm kernel can also neutralize excess acidic substances in the rubber layer, prevent excessive decomposition of the cured rubber layer, and improve the aging resistance of urea-formaldehyde and melamine-urea-formaldehyde resin adhesives [13, 14]. It was also concluded that the stability and initial viscosity of a resin, its pre-compression behavior, and the bonding strength of adhesive products were improved. Walnut shell flour is a filler that is incorporated in urea or resorcinol adhesives to improve spreading or reduce penetration into open wood pores [15–17]. In addition, sorghum flour, protein, bark, and lignin, these kinds of agricultural, forestry, and industrial wastes as fillers have been used in plywood adhesives system [18–21].

In general, different fillers have different advantages. Although kaolin has better properties than flour and calcium carbonate, flour is renewable and sustainable. Most importantly, it is much cheaper, resulting in broader applications in the wood panel industry.

5. Conclusions

Fillers are low-cost additives for wood adhesives during the manufacture of wood composites. They undergo no chemical reactions with the components of wood adhesive systems and can improve some properties or even provide new functions. In the wood panel industry, almost all factories use kaolin clay or flour blended with other components in wood adhesive systems to reduce undesired flow and overpenetration into wood pores in the glue interphase. With the development of society, low-carbon economy, energy conservation, and environmental issues will drive future adhesive developments. Thus, we can predict that future fillers will be functional, differentiated, refined, nanosized, dust-free, and environmentally benign.

Acknowledgements

This work is supported by the following grants and programs: (1) National Natural Science Foundation of China (NSFC 31971595, 31760187); (2) Yunnan Provincial Applied and Basic Research Grants (2017FB060); (3) "Ten-thousand Program"-youth talent support program, and (4) Yunnan Provincial Reserve Talents for Middle & Young Academic and Technical Leaders.

Conflict of interest

There is no conflict of interest in this field.

Author details

Long Cao†, Xiaojian Zhou*† and Guanben Du*
Yunnan Provincial Key Laboratory of Wood Adhesives and Glued Products,
Southwest Forestry University, Kunming, China

*Address all correspondence to: xiaojianzhou@hotmail.com
and gongben9@hotmail.com

† These authors are contributed equally to this work.

References

[1] Dunky M. Adhesives in the wood industry. Handbook of Adhesive Technology. 2003;**2**:50

[2] Frihart CR. Wood Adhesion and Adhesives. Boca Raton, FL: CRC Press; 2005

[3] Dunky M, Pizzi A. Wood adhesives. Adhesion Science and Engineering. Elsevier Science BV; 2002. pp. 1039-1103

[4] Qiao L, Easteal AJ, Bolt CJ, et al. The effects of filler materials on poly (vinyl acetate) emulsion wood adhesives. Pigment & Resin Technology. 1999;**28**(6):326-330

[5] Lei H, Du G, Pizzi A, Celzard A. Influence of nanoclay on urea-formaldehyde resins for wood adhesives and its model. Journal of Applied Polymer Science. 2008;**109**(4):2442-2451

[6] Zhou X, Pizzi A, Du G. The effect of nanoclay on melamine-urea-formaldehyde wood adhesives. Journal of Adhesion Science and Technology. 2012;**26**(10-11):1341-1348

[7] Ghosh PK, Patel A, Kumar K. Adhesive joining of copper using nano-filler composite adhesive. Polymer. 2016;**87**:159-169

[8] Liu D, Chen H, Chang PR, et al. Biomimetic soy protein nanocomposites with calcium carbonate crystalline arrays for use as wood adhesive. Bioresource Technology. 2010;**101**(15):6235-6241

[9] Zhang B, Chang Z, Li J, et al. Effect of kaolin content on the performances of kaolin-hybridized soybean meal-based adhesives for wood composites. Composites Part B: Engineering. 2019;**173**:106919

[10] Wang Z, Li Z, Gu Z, et al. Preparation, characterization and properties of starch-based wood adhesive. Carbohydrate Polymers. 2012;**88**(2):699-706

[11] Zhang Y, Ding L, Gu J, et al. Preparation and properties of a starch-based wood adhesive with high bonding strength and water resistance. Carbohydrate Polymers. 2015;**115**:32-37

[12] Imam SH, Gordon SH, Mao L, et al. Environmentally friendly wood adhesive from a renewable plant polymer: Characteristics and optimization. Polymer Degradation and Stability. 2001;**73**(3):529-533

[13] Zhao X, Peng L, Wang H, et al. Environment-friendly urea-oxidized starch adhesive with zero formaldehyde-emission. Carbohydrate Polymers. 2018;**181**:1112-1118

[14] Ong HR, Khan MMR, Prasad DMR, et al. Palm kernel meal as a melamine urea formaldehyde adhesive filler for plywood applications. International Journal of Adhesion and Adhesives. 2018;**85**:8-14

[15] Khanjanzadeh H, Pirayesh H, Sepahvand S. Influence of walnut shell as filler on mechanical and physical properties of MDF improved by nano-SiO 2. Journal of the Indian Academy of Wood Science. 2014;**11**(1):15-20

[16] Pizzi A. Resorcinol adhesives. Handbook of Adhesive Technology. Taylor & Francis Group, LLC; 2003:599-613

[17] Zhou X, Pizzi A. Pine tannin based adhesive mixes for plywood. International Wood Products Journal. 2014;**5**(1):27-32

[18] Hojilla-Evangelista MP, Bean SR. Evaluation of sorghum flour as extender in plywood adhesives for sprayline coaters or foam extrusion.

Industrial Crops and Products. 2011;**34**(1):1168-1172

[19] Kang H, Wang Z, Wang Y, et al. Development of mainly plant protein-derived plywood bioadhesives via soy protein isolate fiber self-reinforced soybean meal composites. Industrial Crops and Products. 2019;**133**:10-17

[20] Aydin I, Demirkir C, Colak S, et al. Utilization of bark flours as additive in plywood manufacturing. European Journal of Wood and Wood Products. 2017;**75**(1):63-69

[21] Olivares M, Aceituno H, Neiman G, et al. Lignin-modified phenolic adhesives for bonding radiata pine plywood. Forest Products Journal. 1995;**45**(1):63

Chapter 3

Fillers for Packaging Applications

Giovani Otavio Rissi

Abstract

Packaging in general is frequently overlooked and demonized. The lack of educational programs and efficient waste treatment lead packaging to be treated as an environmental problem. However, packaging is an enabler of our society because it makes feasible the availability of any and every good, regardless of its production location. Furthermore, the packaging business plays a significant role in the global economy, following a continuous trend of growth. The use of fillers in various packaging types can be a valuable resource not only for reducing its cost but also improving its mechanical strength (therefore reducing the number of raw materials required for making that specific package), improving its visual properties to ensure customer attractiveness, creating new possibilities of use, and extending the shelf life of perishable foods. However, the use of fillers in packaging should be made in a way that permits proper recovery and recycling after use.

Keywords: packaging, packaging market, paper packaging, plastic packaging, nanofillers, circular economy

1. Introduction

Before proceeding, it is worthwhile to make a little observation about the environment around you. Take a quick break and behold what surrounds you and the place where you are.

Unless you are reading this in the wild of nature, everything you see was made available through some sort of packaging, even the materials used in the construction of the place where you are were bundled, contained, or packed.

Also, everything you eat—regardless of being an ordinary meal or a gourmet delight—came to you via packaging entrusted to keep high food safety and hygiene standards. Even if one has a backyard or farm that provides a wide variety of fresh produce, at some point in time some packaged foods will be bought and brought home.

Now think about all medicines that are consumed by millions of people every day to keep their well-being and health. Without being noticed, packaging allows for one of the most noble uses: to provide a longer and healthier life. You may now have realized for the first time that life as we know it exists at current standards due to a powerful enabler: packaging. Some studies suggest an association between the quality of life in a certain location and the level of packaging development in the same place [1].

The reason for this is simple: our society has developed in a way that knowledge and experimentation are part of everyday life. We are informed about the latest trends and we want to try new things. At the same time, just a small fraction of the goods we buy is produced in our vicinity. Most goods are produced thousands of kilometers away from where we live. Some make journeys to our home longer than the longest travel we have ever done.

The definition of the term "packaging" is very broad [2, 3], but it is possible to define it as "a system that consists of different materials, machinery and symbols which are set up in a way to contain, protect, communicate and allow access of its contents to make goods available in a cost-efficient way."

It is not the intention of this chapter to go into detail on marketing aspects of packaging, but references will be made when necessary since most products on shelves rely on their packaging as the sole communication tool with their consumers.

2. Packaging market

The global market of packaging was estimated to be as of US$ 851 billion in 2018 as per recent estimates (**Figure 1**) [4, 5].

Considering the market share by material types, cardboard (including corrugated boxes) and plastic packaging in their flexible and rigid forms correspond to almost 80% of the whole market. Fillers are used mainly in these three types of packaging.

Cardboard production has increased due to the rapid growth of e-commerce in recent years. Flexible packaging has proven to be a lightweight and cheaper alternative to glass and metal. The annual growth of the packaging segment is estimated to be between 3.5 and 4.3% and is influenced by the urban shift of populations, the trend of pre-packed products in the food segment, and diversification of retail landscape [6].

It is important to highlight that these estimates were made before the COVID-19 pandemic of 2020, economy was still not precise by the time this chapter was being written. Although exact figures cannot be given, packaging market growth is expected to follow global trade trends, according to previous studies [7].

It is estimated that in 2018 the global plastic production almost reached 359 million tonnes. Even in Europe, where environmental concerns are usually high and public opinion plays a big role, packaging is of key importance since it represents almost 40% of the plastics demand. This share is very expressive since it is more

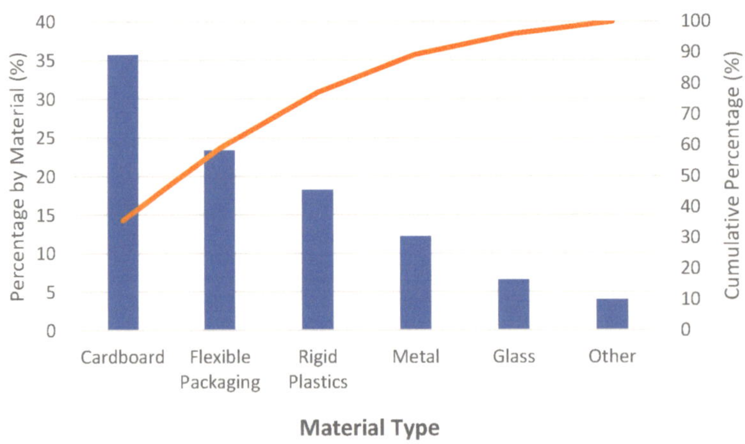

Figure 1.
World packaging market (share by material) [4, 5].

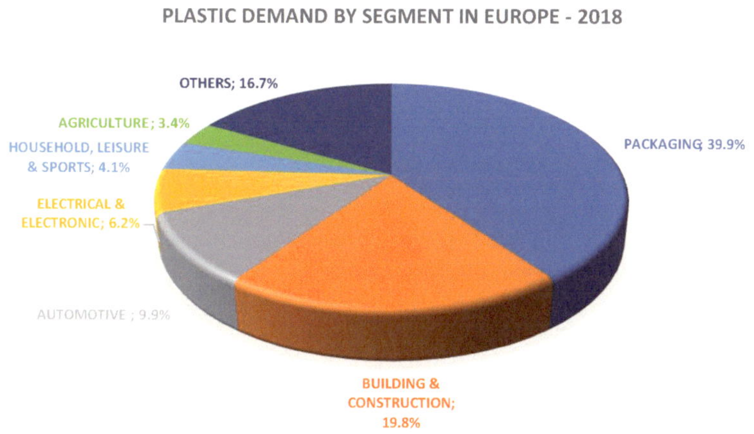

Figure 2.
Plastic uses in Europe [8].

than double the second main use of plastics (building and construction, which corresponds to almost 20% of share) (**Figure 2**) [8].

3. Fillers used in paper packaging

There are two major categories of cellulose-base packaging: corrugated cases—the standard brown box using for transporting—and paperboard carton as you may see on supermarket shelves containing cereals, toothpaste, frozen food, etc. There is a wide array of drawing possibilities in both categories and the choice about what should be used depends on product requirements, machinability, and design strategy. These materials are used for "secondary packaging" (because it does not get into contact with the product) in the case of paperboard carton or "transport packaging" in the case of corrugated cases for obvious reasons.

The enhancement of visual properties in paper packaging given by fillers makes its use imperative for assuring customer attractiveness. This is not the case for corrugated cases, which have the function of protecting and grouping various units of products during transport. However, for paperboard carton packaging, this feature is crucial for standing out on shelves, grabbing consumer's attention, and providing improved sales.

Fillers are not added to the process of "converting" cellulosic packaging (printing, cutting, creasing, gluing, etc.) but in the production of the various paper grades that will be further transformed.

Fillers used in papermaking are defined as insoluble particles (from 0.1 to 10 μm) added to the slurries of cellulosic fibers before the formation of paper. The uses of fillers vary from zero to 30% [9]. On average, it is estimated that minerals used for wet-end addition and coatings make up 8% (by mass) of the materials used in the paper industry [10].

What defines the amount of fillers used in papermaking is the grade and purpose of that specific paper or paperboard. For instance: an aseptic multilayer packaging like the milk box may have little or no fillers at all, while printing papers may have 30% (by mass) or more of its content made by fillers.

There has been a noted increase in the number of fillers being used, driven by lower cost of fillers themselves compared to the fibers itself, and by meeting market demand for higher opacity and/or brightness. The most common fillers found in papermaking can be seen in **Table 1** [11].

Chemical composition	Natural source	Synthetic source
$CaCO_3$	Ground limestone (GCC) Chalk (ground)	Precipitated calcium carbonate (PCC)
$Al_2O_3.2SiO_2.2H_2O$	Clay (hydrous kaolinite)	Precipitated aluminum silicate
TiO_2	—	Titanium dioxide
$Mg_3Si_4O_{10}(OH)_2$	Talc	—
$CaSO_4.2H_2O$	—	Gypsum
Adapted from [11].		

Table 1.
Common fillers used in papermaking.

Fillers from renewable sources can also be used in papermaking [12]. Most common types of natural fillers are:

- **Starch or starch derivates**—for increasing fiber bonding and consequently mechanical properties like tensile, tearing and folding strength.

- **Cellulose derivates (like microcrystalline cellulose)**—for encapsulating fillers and increasing its bonding strength to cellulose fibers, as well as improving heat capacity.

- **Chitin or chitosan**—also for improving bonding, since chemically speaking the structure of the molecules of chitin and chitosan are very similar to that of cellulose.

- **Xantham gum or anionic guar gum**—for improving paper strength.

Although the use of fillers can influence a wide array of attributes in papermaking (coefficient of friction, permeability, burn rate), the two main drivers for their use in cellulose-based packaging are:

- **Cost reduction**—this can be achieved in two ways: either by partially substituting cellulose content or by reducing the energy demand required for drying paper.

- **Visual properties**—the addition of fillers greatly enhances printing quality on paper packaging by improving opacity, brightness, and smoother surface from calendaring. This is particularly important for products that communicate their brand through their shelf packaging, with the need of attracting customers.

Other uses of fillers can lead to the development of nonconventional applications, although commercial use of several novel applications is still strict [13]. Examples are deodorant paper [14], antimicrobial paper [15], flame retardants [16], and magnetic paper [17].

However, fillers reduce the flexibility of paper, posing extra attention to converting operations. Side effects, like creasing, if not done properly the packaging may show small cracks when folded, resulting in reduced mechanical strength and visual defects that can compromise the packaging function of protection and its attractiveness at the point of sale. Furthermore, and due to its mineral nature, fillers increase abrasion and dusting.

To overcome these drawbacks, fillers have been modified in several ways to achieve desired properties in an optimized way. Examples of processes that allow filler modification are modification with inorganic substances, modification with natural polymers (or their derivatives), modification with water-soluble synthetic polymers, modification with surfactants, hydrophobic modification, cationic modification among others [13].

4. Fillers in plastic packaging

Plastic packaging is widely used due to its wide range of applications, low cost, relatively high level of safety in terms of migration, and the convenience it can add to packaged products. They have been extensively employed for substituting more expensive packaging materials like glass, metal, and even wood [18].

Plastics are commonly used as "primary packaging," meaning it directly contacts the product. Plastics can also be used as secondary packaging to bundle bottles for instance or to transport packaging in the form of a stretch film.

Regardless of its use, there are two major groups of plastic packaging, based on their physical characteristics:

- Flexibles—which include all wraps, bags, and other packaging that can be used manually or automated through vertical or horizontal machines.

- Rigid—which includes all sorts of tubes, bottles, containers, and drums. This type of packaging is suitable for liquids and products requiring mechanical protection or specific positioning. Due to the higher amount of mass required for achieving rigidity, it is also more expensive and substituted with flexible alternatives whenever possible.

Recent developments in the area of material engineering make use of a wide variety of fillers to modify the original properties of plastics. Advances in the production of nanofillers also have shown promising areas of research. These functional fillers can result in unique properties that enhance the performance of packaging materials even at small loads.

In theory, any material that can be found or transformed into small particles can be used as a filler for plastics in general. Potential combinations are endless but for commercial purposes, fillers for plastics should be readily available, insoluble, and chemically inert; have a low hardness to avoid wear; and be nontoxic, non-flammable, and finally free from metal impurities, which can degrade plastics [19].

In general, the most common fillers used in plastics are calcium carbonate, aluminum trihydrate, talc, kaolin, mica, wollastonite, glass fiber, aramid fiber, carbon fiber, and carbon black. However, it is important to highlight that fillers are used mostly in engineered plastics (**Table 2**).

The use of fillers affects plastics by modifying original material properties or adding features that did not exist in the original polymer. Specifically for this aim, properties can be changed due to the use of glass fibers, mica flakes, nano-clays, carbon nanotubes/nanofibers, natural fibers, wood flour, talc, and kaolin [21].

Plastic packaging containing nanofillers has been shown to have an enhanced performance especially for food packaging, where high barrier properties against water vapor, aromas, and oxygen are mandatory [22].

Filler source	Group	Examples
Inorganic	Oxides	Glass, SiO_2, ZnO, Al_2O_3, MgO
	Hydroxides	$Mg(OH)_2$, $Al(OH)_3$
	Salts	$CaCO_3$, $CaSO_4$, $BaSO_4$, phosphates
	Silicates	Talc, kaolin, mica, montmorillonite, wollastonite, feldspar
	Metal	Steel, boron, (silver for antimicrobial nanofillers)
Organics	Carbon, graphite	Carbon fillers and nanotubes, carbon black, graphite fibers and flakes
	Natural polymers	Cellulose, starch
	Synthetic polymers	Polyester, aramid, polyamide and PVOH fibers
Natural fibers	Straw	Wheat, corn, and rice
	Bast	Hemp, jute, kenaf, lax
	Leaf	Pineapple leaf, sisal
	Seed/fruit	Cotton, coir
	Grass fibers	China reed, bamboo, grass
	Wood fiber	Hardwood, softwood

Adapted from [20, 21].

Table 2.
Examples of fillers used in plastics.

Because the surface area of nano-clays can be more than 750 m^2/g, the use of nanofillers even in loads smaller than 2% (in volume) creates a tortuous path for the diffusion of gases through the polymer matrix, therefore improving its gas barrier. This is of special interest because nanofillers can be an alternative to multilayer coextruded packaging, which are commonly not recyclable. Another advantage of adding nanofillers (like montmorillonite-MMT, kaolinite, carbon nanotubes, and more recently graphene nanosheets) in plastic packaging is the improved mechanical properties that allow downgauging and consequent economic and environmental benefits.

The use of nanofillers can also enable broader use of biopolymers in packaging. This group of materials has been studied for several decades but, usually, their properties are much weaker when compared to petroleum-based polymers. They include natural sources like starch; cellulose; proteins (collagen, soybean protein, zein, etc.); and polylactic acid (PLA). The low-performance properties of these "green" materials and can be compensated by the addition of nanofillers, broadening the possibilities of use at commercial scale [23, 24].

On top of the mechanical reinforcement and improved barrier properties, nanofillers can extend the shelf life of products that may spoil due to the development of pathogenic or spoilage microorganisms. It is known that the use of nanofillers made of silica through the sol-gel method has improved the shelf life of fruits due to their hydrophobic characteristics [25].

Besides the features mentioned above, other beneficial functions of nanofillers in food packaging are the possibility of exploring and developing new technologies in the area of active packaging, where the role of packaging goes beyond traditional purposes. Main groups of research areas can be listed as follows [26, 27]:

a. Oxygen scavenging—oxygen causes food spoilage due to oxidation or due to enabling aerobic bacteria development. Even under proper vacuum, package materials may allow the permeation of O_2.

b. Nanobased sensors—for compensating transport and storage conditions (e.g., temperature variations) and informing about the potential shorter shelf life of products.

c. Detection of gases produced by food spoilage—microbial development is commonly followed by gas production, which can be detected by conducting polymer nanocomposites.

d. O_2 indicators—since the development of aerobic microorganisms happens during the presence of oxygen, the control of this gas is necessary to evaluate product decay.

e. Enzyme immobilization at the nanoscale—sometimes the direct uses of enzymes is restricted due to its potential degradation during processing. The absorption of enzymes by nano-clays embedded in a polymer matrix is a promising mechanism for efficient release control.

However, the fillers at nanoscale should be only used after a careful study of their toxicology and potential harm to humans and the environment. It is known that migration may happen from food-contact materials through various mechanisms that may not be fully extracted by standardized simulants or quantified by current analytical methods [28].

5. Environmental concerns

More than ever, the packaging is under public scrutiny and blamed for environmental problems. Although only approximately 10%, of all solid waste, is plastic, up to 80% of all waste accumulated in land, shorelines, seabed, and ocean surface is plastic due to its long decomposition time. This is a threat to aquatic animals since they can be entangled or ingest plastic fragments. Some researches suggest that over 260 species of animals (invertebrates, mammals, seabirds, fishes, turtles) have ingested or been entangled by plastic debris.

Due to its fragmentation properties, plastic packaging can decompose into microplastics and pose yet another type of hazard. The bodies of all marine species, ranging in size from plankton to the blue whale, contain plastics [29]. The food chain has detrimentally changed.

While attention has been focused on the number of microplastics in oceans, almost no effort has been made in fighting root causes. The pollution on land, in rivers and the ocean, is not caused by plastic packaging or any other packaging material, but by the lack of educating the public on the consequences of littering.

Furthermore, massive marketing is promoting miraculous solutions and materials for the sake of the environment without technical evidence about its effectiveness. Deceiving customers with the promise of mitigating their environmental impact is called greenwashing.

Another problem in evaluating the environmental benefit of packaging is to neglect the components of the product itself. It happens for the vast majority of products that the packaging impact is much lower than that of the product itself [30] and eventual packaging failure will lead to increased waste.

The most advanced tool to evaluate the eco-efficiency of packaging is the Life Cycle Assessment (LCA), which has already been used for a wide array of applications with great acceptance from the scientific community. It provides comparative data about the environmental impact of the product under analysis considering various aspects like Global Warmth Power (GWP), ozone depletion, toxicity (both cancerous and noncancerous), particulate respiratory effects, ionizing radiation, photochemical ozone creation potential, acidification potential, aquatic eutrophication, freshwater ecotoxicity potential, and nonrenewable resource depletion. Other parameters to be considered are land use, freshwater use, and cumulative energy demand (CED) [31].

It was already mentioned that fillers reduce the drying energy required in papermaking, which will be transformed into cellulosic-based packaging. Although the use of fillers may reduce the carbon footprint of papermaking in some aspects [32], a full LCA analysis should be performed to confirm assumptions. Nevertheless, the environmental impact of mineral fillers in the production of paper or paperboard is expected to be negative according to perspectives below:

a. **Substitution of cellulosic fibers**—during plant growth, CO_2 from the atmosphere is captured resulting in negative emissions. Although industrial processes have positive emissions for the production of paper, the net absorption of CO_2 is much larger. Numbers vary according to countries, paper grades, and technological development, but public data from companies that have an integrated paper facility show that every ton of finished product captures the equivalent of 2.6–3.8 tons of CO_2 equivalent ($CO_{2eq.}$) from the atmosphere [33, 34]. Adding fillers for the sake of sheer cost reduction with no technical benefits increases the environmental impact of the final packaging (in terms of $CO_{2eq.}$).

b. **Replacement of a renewable raw material by a nonrenewable one**—although synthetic fillers can be produced, mineral fillers of natural sources are extensively used. They are mined in several parts of the world, which means dependence on the natural availability of that specific resource. On the other hand, cellulose fibers come from trees, which can be planted and harvested repeatedly, if responsible soil and water handling are provided.

c. **Harder to recycle**—many cellulosic-based packaging may be produced from recycling another packaging due to availability and costs. If the paperboard can not be easily recycled, it is likely to be sent to energy recovery (incineration) or, more commonly, landfilling. During the degradation of cellulose, CH_4 is emitted, representing a GWP of about 25 times that of CO_2 [35]. Depending on the quality and grade of the recycled paper pursued, fillers have to be removed to provide a higher quality of recycled materials (which also have a higher market value). It is estimated that the removal of fillers along with other contaminants (stickers, inks, fines, etc.) require 30–100% more recovered paper for the same final product, resulting in a higher amount of nonrecyclable waste to deal with [36].

d. **Environmental impact for making fillers**—the process of making fillers comprehends an extensive production chain, which includes removal of impurities, engineering particle shape and size, and finally enhancing its properties through thermal and/or chemical treatments. Also, fillers should be transported to the paper mill facility, which may be located far away, incurring fuel consumption and extra environmental burden. The mining activity alone is very complex: it includes exploration of the deposits, resource development, feasibility and reserve development, mine planning (including permitting and construction), and production. When resources are exhausted and the mine is closed, further work is necessary for restoring the area [37].

Fillers of organic origin like wood waste-derived fillers have been studied for several decades and can contribute toward reducing solid waste, even in large-scale commercial activities, although compounders are somehow reluctant to consider this technology because they are not sure how to handle it or are not aware of a market to justify production [38, 39].

Unfortunately, for the time being, the amount of LCA studies evaluating the environmental impact of fillers added to polymers is likewise small but indicates potential lower GWP comparing to standard plastics. Better eco-performance is achieved using nanofillers, which reduces the amount of petroleum-based raw materials for achieving the same performance although some exceptions can be found (e.g., LDPE and talc).

The addition of natural fibers to partially substitute polymers may also be an opportunity for reducing the environmental impact especially because they come from renewable sources, but again an LCA analysis is necessary for confirmation due to the production burden of raw materials [40]. It is important to highlight that LCA studies should include not only the resources used to produce that specific type of packaging but also the after-use impact. This broad view limits the boundaries of the analyses "from cradle to grave" and is fundamental for having an overall picture of the system.

The process of recovering material by recycling is becoming increasingly important due to environmental and economical concerns. One recent driver toward reduced environmental impact is the concept of circular economy, where the traditional linear model of production is substituted by the concept of keeping goods in use for the longest possible time (**Figure 3**) [41].

One of the models of circular economy in use was proposed by the Ellen MacArthur Foundation, commonly known as the "butterfly diagram." The smaller the loop, the greater the value of the material in question. The right side of the diagram (in blue) refers to "technical cycles" where goods are designed to stay in use for the longest possible period through sharing, reusing, extended durability, repairing, and recycling, the last possible loop.

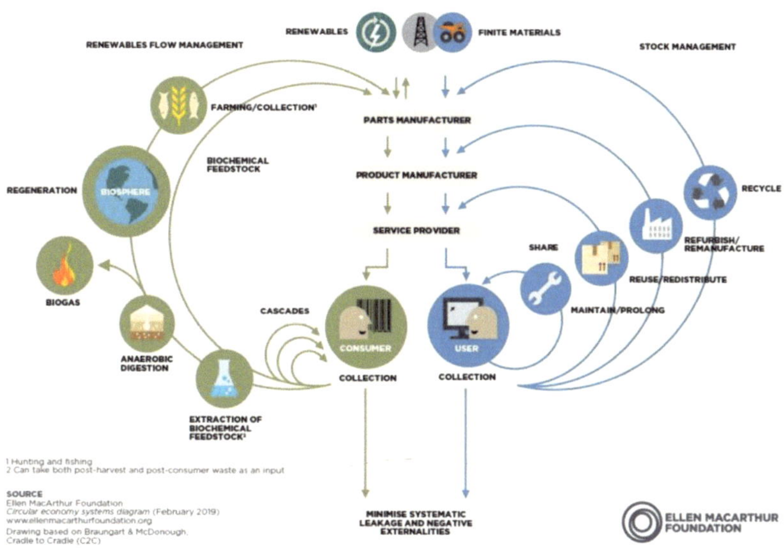

Figure 3.
The circular economy model with technical (right) and biological (left) cycles proposed by the Ellen MacArthur Foundation. Reproduced with permission.

The left part of the diagram (in green) refers to "biological cycles," where components of any goods or packaging may biodegrade. This is a feasible alternative for organic waste or sewage, but it is the least wanted solution for packaging because recycling has a greater value and lower environmental impact than biodegrading.

6. Conclusion

Packaging plays a fundamental role in our society by providing accessibility to food and any other type of physical goods. Broad multidisciplinary knowledge is necessary for making the definition of a packaging system accurately.

The choice of packaging which is more adequate for a specific product must pass through the definition of suitable materials and its properties for best fulfilling production, distribution, preservation and sales requirements. Most important packaging materials are made of cellulose (cases and folded carton) and polymers (flexible or rigid packaging), which are the types of materials where fillers are mostly used.

No matter what type of packaging system is chosen, the end of life of that specific packaging (reuse, recycling, incineration or composting) should always be considered in the project phase. The selection of the proper filler(s) combined with the optimum amount used to contribute to the success of that package either during its lifespan as well as during the recycling stage.

Acknowledgements

The author would like to thank Juliano Barbosa and Stefan Viering for sharing their expertise. The author is also grateful to Hessa Tary for her feedback and friendship.

Conflict of interest

The author declares no conflict of interest.

Author details

Giovani Otavio Rissi
Independent Packaging Consultant, Gavardo, Italy

*Address all correspondence to: giovani.rissi@gmail.com

References

[1] Bülent E, Aysel I. Packaging materials and effects on quality of life. In: Proceedings of the 1st International conference on Quality of Life; June 2016. Serbia: Center for Quality, Faculty of Engineering, University of Kragujevac; 2016

[2] Emblem A, Emblem H, editors. Packaging Technology: Fundamentals, Materials and Processes. Cambridge: Woodhead Publishing; 2012. p. 579

[3] Hanlon F.J, Kelsey R. J, Forcinio, H, authors. Handbook of Package Engineering, 3rd ed. Boca Raton: CRC Press; 1998. 698

[4] The Future of Global Packaging to 2022. Smithers Pira [Internet]. 2018. Available from: https://www.smithers.com/resources/2018/feb/paper-and-boardroles-in-the-future-of-packaging [Accessed: 15 April 2020]

[5] Packaging Market Continuous to Grow [Internet]. Available from: https://www.interpack.com/en/TIGHTLY_PACKED/SECTORS/INDUSTRIAL_GOODS_PACKAGING/News/Packaging_market_continues_to_grow [Accessed: 15 April 2020]

[6] Packaging and Protective Packaging [Internet]. 2020. Available from: https://www.alliedmarketresearch.com/packaging-and-protective-packagingmarket [Accessed: 17 April 2020]

[7] Packaging and Trade Kofi Manso Essuman [Internet]. 2018. Available from: https://www.worldpackaging.org/Uploads/2018-11/ResourcePDF27.pdf [Accessed: 22 April 2020]

[8] Plastics—The facts 2019 [Internet]. 2019. Available from: https://www.plasticseurope.org/application/files/9715/7129/9584/FINAL_web_version_Plastics_the_facts2019_14102019.pdf [Accessed: 22 April 2020]

[9] Hubbe M, Gill R. Fillers for papermaking: A review of their properties, usage practices, and their mechanistic role. BioResources. 2016;**11**:2886-2963. DOI: 10.15376/biores.11.1.2886-2963

[10] Chamberlain D, Kirwan M. Chapter 1 - Paper and paperboard - Raw materials, processing and properties. In: Handbook of Paper and Paperboard Packaging Technology. 2nd ed. Sussex Willey-Blackwell; 2012. pp. 1-49. DOI: 10.1002/9781118470930.ch1

[11] Gill RA. Applications of wet-end paper chemistry fillers for papermaking. In: Au CO, Thorn I, editors. Applications of Wet-End Paper Chemistry. Dordrech: Springer; 1995. pp. 54-75. DOI: 10.1007/978-94-017-0756-5_4

[12] Shen J, Song Z, Qian X, Ni Y. Carbohydrate-based fillers and pigments for papermaking: A review. Carbohydrate Polymers. 2011;**85**:17-22. DOI: 10.1016/j.carbpol.2011.02.026

[13] Shen J, Song Z, Qian X, Ni Y. Modification of papermaking grade fillers: A brief review. BioResources. 2009;**4**:1190-1209

[14] European Patent Specification, A functional Paper and Its Use as a Deodorant, Filtering Medium or Adsorbent [Internet]. 1997. Available from: https://patentimages.storage.googleapis.com/ee/5e/5b/f254362b546ee8/EP0393723B1.pdf [Accessed: 27 April 2020]

[15] Kim C-H, Cho S-H, Park W-PO. Inhibitory effect of functional packaging paper containing grapefruit seed extracts and zeolite against microbial growth. Appita Journal. 2005;**58**:202-207

[16] United States Patent—Fire retardant pigment [Internet]. 1989. Available from: https://patentimages.storage.googleapis.com/17/b2/a4/456a42fe06b828/US4842772.pdf

[17] Zakaria S, Ong BH, van de Ven T. Lumen loading magnetic paper I: Flocculation. Colloids and Surfaces A: Physicochemical and Engineering Aspects. 2004;**251**:1-4. DOI: 10.1016/j.colsurfa.2004.06.020

[18] DeArmitt C. Functional fillers for plastics. In: Kutz M, editor. Applied Plastics Engineering Handbook: Processing and Materials. Oxford: Elsevier; 2011. pp. 455-468. DOI: 10.1016/B978-1-4377-3514-7.10026-1

[19] Filled with Success [Internet]. 2010. Available from: https://phantomplastics.com/wp-content/uploads/2013/08/Filled-with-Success-Composites-World-July-August-2010.pdf [Accessed: 05 May 2010]

[20] Bos HL. The Potential of Flax Fibres as Reinforcement for Composite Materials. Eindhoven: Technische Universiteit Eindhoven; 2004. DOI: 10.6100/IR575360

[21] Xanthos M, editor. Polymers and polymer composites. In: Functional Fillers for Plastics. Weinheim: WILEY-VCH Verlag GmbH & Co. KGaA; 2011. p. 451. DOI: 10.1002/9783527629848.ch1

[22] Ahmed MR. Polymer nanocomposites as a new trend for packaging applications. Polymer-Plastics Technology and Engineering. 2013;**52**(7):635-660. DOI: 10.1080/03602559.2012.762673

[23] Arora A, Padua GW. Review: Nanocomposites in food packaging. Journal of Food Science. 2010;**75**:R43–R49. DOI: 10.1111/j.1750-3841.2009.01456.x

[24] Chielini E. Environmentally Compatible Food Packaging. Cambridge: Woodhead Publishing; 2008. p. 592

[25] Arreche R, Blanco M, Vázquez P, Martin-Martinez J. Use of new silica fillers as additives for polymers used in packaging of fruit. Química Nova. 2011;**35**:1907-1911. DOI: 10.1590/S0100-40422012001000003

[26] Azeredo H. Nanocomposites for food packaging applications. Food Research International. 2009;**42**(9):1240-1253. DOI: 10.1016/j.foodres.2009.03.019

[27] Azeredo H, Otoni C, Assis O, Correa D, Moura M, Mattoso L. Nanoparticles and Antimicrobial Food Packaging. Basel MDPI. 2018. DOI: 10.1016/B978-0-08-1005965.21874-X

[28] Maryam J, Gitte AP, Katrin L. Six open questions about the migration of engineered nano-objects from polymer-based food-contact materials: A review. Food Additives & Contaminants: Part A. 2017;**34**(3):434-450. DOI: 10.1080/19440049.2016.1271462

[29] Wabnitz C, Nichols W. Editorial: Plastic pollution: An ocean emergency. Marine Turtle News Letter. 2010;**129**:1-4

[30] Grant T, Barichello V, Fitzpatrick L. Accounting the impacts of waste product in package design. In: Procedia CIRP; 7-9 April 2015; Sydney, Australia. p. 568572. DOI: 10.1016/j.procir.2015.02.062

[31] Tacker M. Methods for the assessment of environmental sustainability of packaging: A review. IJRDO. 2017;**3**:33

[32] Shen J, Song Z, Qian X. Fillers and the carbon footprint of papermaking. BioResources. 2010;**5**:2026-2028. DOI: 10.15376/biores.5.4.2026-2028

[33] Relatório de sustentabilidade Irani [Internet]. 2018. Available from:

http://www.irani.com.br/wp-content/uploads/2019/04/Relat%C3%B3rio-de-Sustentabilidade-Irani-2018.pdf [Accessed: 16 May 2020]

[34] Inventário Corporativo de Gases de Efeito Estufa [Internet]. 2013. Available from: http://www.irani.com.br/uploads/mediacenter/9dafb35a11fe587a8b093f2b76871241f823eda6.pdf [Accessed: 16 May 2020]

[35] Ongmongkolkul A. Life cycle assessment of paperboard packaging produced in Thailand [M.S. thesis]. Thailand: School of Environment, Resource and Development Asian Institute of Technology (AIT); 2001

[36] Ghinea C, Petraru M, Simion I, Sobariu D, Bressers H, Gavrilescu M. Life cycle assessment of waste management and recycled paper systems. Environmental Engineering and Management Journal. 2014;**13**:2073-2085. DOI: 10.30638/eemj.2014.230

[37] Kogel JE. Mining and processing kaolin. Elements. 2014;**10**(3):189-193. DOI: 10.2113/gselements.10.3.189

[38] English B, Clemons CM, Stark N, Schneider JP. Waste-wood-derived fillers for plastics. In: Gen. Tech. Rep. FPL-GTR-91. Madison, WI: U.S. Department of Agriculture, Forest Service, Forest Products Laboratory; 1996. p. 15. DOI: 10.2737/FPL-GTR-91

[39] Clemons CM. Natural fibers. In: Xanthos M, editor. Functional Fillers for Plastics. Weinheim, WILEY-VCH Verlag GmbH & Co. KGaA. 2010. DOI: 10.1002/9783527629848.ch11

[40] Civancik-Uslu D, Ferrer L, Puig R, Fullana-i-Palmer P. Are functional fillers improving environmental behavior of plastics? A review on LCA studies. Science of the Total Environment. 2018;**626**:927-940. DOI: 10.1016/j.scitotenv.2018.01.149

[41] The Circular Economy in Detail [Internet]. 2017. Available from: https://www.ellenmacarthurfoundation.org/explore/the-circular-economy-indetailhttps://www.ellenmacarthurfoundation.org/explore/the-circular-economy-in-detail [Accessed: 15 May 2020]

Chapter 4

Towards Traditional Carbon Fillers: Biochar-Based Reinforced Plastic

Mattia Bartoli, Mauro Giorcelli,
Pravin Jagdale and Massimo Rovere

Abstract

The global market of carbon-reinforced plastic represents one of the largest economic platforms. This sector is dominated by carbon black (CB) produced from traditional oil industry. Recently, high technological fillers such as carbon fibres or nanostructured carbon (i.e. carbon nanotubes, graphene, graphene oxide) fillers have tried to exploit their potential but without economic success. So, in this chapter we are going to analyse the use of an unconventional carbon filler called biochar. Biochar is the solid residue of pyrolysis and can be a solid and sustainable replacement for traditional and expensive fillers. In this chapter, we will provide overview of the last advancement in the use of biochar as filler for the production of reinforced plastics.

Keywords: biochar, composites, sustainable production, carbon materials

1. Introduction

Carbon-based materials are a very well-established commodity generally used in materials science [1]. Nowadays, many commodities makes use of carbon fibres as they become an unavoidable asset for the global market [2]. Carbon fibres represent the most diffuse high-tech carbon materials, but carbon black played the main role in these materials. Carbon black (CB) harvests a great global carbon revenue due to its use for the production of plenty of composites but mainly for tyres [3]. Over the years, high-cost carbon materials such as carbon nanotubes (CNTs) and graphene-like materials have gained the attention of the scientific community with their amazing conductivity and optic and mechanical properties [4, 5]. Despite the expected revolution, nanosized allotropic carbon form did not have much progress in the research area. In a very optimistic report, Segal [6] dreamed that the world was ready for the industrial-scale production of graphene, but after a decade, single-layer graphene is still sold at 200 €/cm^2, while graphene oxide costs 100,000 €/kg [7]. On the other hand, CB is sold for around 1 €/kg [8]. New-generation high-tech carbon materials (i.e. CNT, graphene and graphene oxide) have not yet fulfilled the promise for a new carbon era. While the industry waits for large-scale commercialization of high-quality affordable carbon allotropes, new materials have been considered through engineered carbon for profitable business. In recent years, a new material has emerged as the most promising for the integration of carbon production with waste management [9–11]. This material is biochar, the solid residue from pyrolytic conversion of biomass. Biomass waste stream is one of the most abundant worldwide, and it is generally disposed through incineration. This presents both an environmental issue and an economic loss due to the transformation of a high-quality material into heat. Accordingly, a more profitable advantage was found in their thermal conversion for the production of biofuels [12, 13], chemicals [14] and other materials [15]. Conversion of biomass into liquid fuels is challenging due to the high oxygen content compared with traditional oil-derived products (i.e. gasoline, virgin nafta, diesel). On the contrary, biochar production is a process full of opportunities with the emergence of carbonaceous material from both lignocellulosic and non-lignocellulosic biomasses. This bioderived carbon is used in many applications [16] due to its properties and low cost attested at around 1–2 €/kg [17–19]. Actually, biochar has found a large-scale application for soil health improvement [20–22] and as solid fuel with a heating content of around 40 kJ/mol [23]. Nonetheless, these applications are limited and unable to exploit the full potentiality of biochar due to their easy tunability with simple process adjustment [24].

In this book chapter, we report an overview of the composite applications of biochar to prove its feasibility as a replacement for traditional carbon materials and as a solid competitor with high-tech reinforced plastics.

2. Biochar: production ways

Biochar is produced through thermochemical routes such as torrefaction, pyrolysis and hydrothermal carbonisation and as residue of gasification.

Torrefaction is a low-temperature thermal conversion used to densify the biomasses for energy purposes [25]. The process temperature is in the range from 200–350°C, and the conversion requires long residence and processing times. Torrefaction is characterised by biochar and biochar-like yields [26]. The carbon percentage of solid residue is generally around 50–60 wt.% [27], but it can reach 72–80 wt.% using microwave process with the addition of microwave absorbers as reported in several studies [28–31]. Microwave use leads to the drastic reduction of process timescale from hours to minutes.

Pyrolysis is a high-temperature thermochemical conversion which induces the cracking of polymers with the formation of low-molecular-weight compounds in an oxygen-free or oxygen-poor atmosphere [12, 32]. Pyrolysis is run using different heating technologies [33] and apparatus design [34–37] at a temperature range from 450–700°C [38] with huge variations in product fraction yields. Pyrolysis of biomasses was deeply studied, and the main mechanisms can be rationalised in a few different steps. The first is the release of moisture from the feedstock, increasing the surface area and improving the pore structure, which favours a quick release of volatiles and minimizes char-catalysed secondary cracking. Lignocellulosic biomass behaviour during pyrolysis could be rationalised through the behaviour of the main components as cellulose, hemicellulose and lignin. The pyrolysis of cellulose takes place between 430 and 470°C and hemicellulose between 470 and 600°C while lignin between 600 and 800°C. During this process other reactions such as dehydration of the sample, pyrolysis of the volatiles present, formation of levoglucosan from cellulose [39] and formation of substituted aromatic rings from lignin [40] take place together with the formation of carbon.

Hydrothermal carbonisation is a thermal cracking used to produce crude-like oil and hydrochar under moderate temperature and high pressure [41] using aqueous solvent [42], nonaqueous solvent [43, 44] or subcritical/critical media [45].

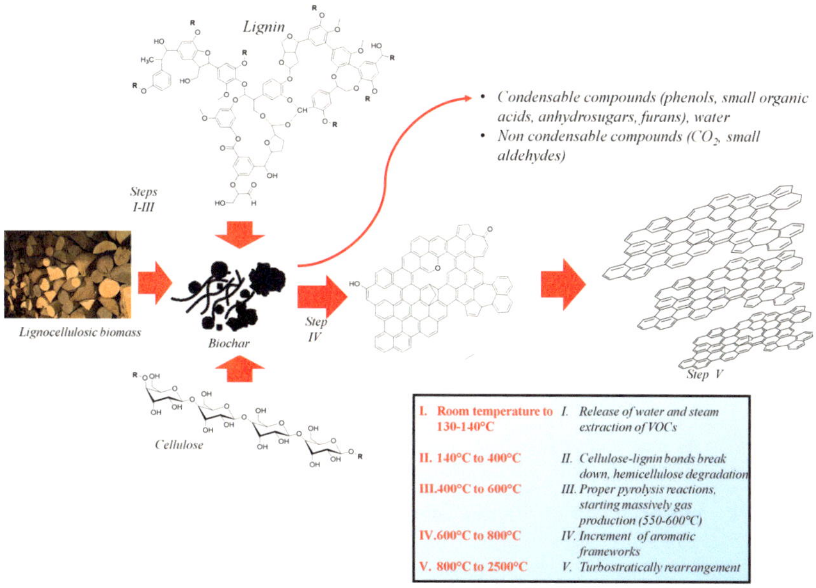

Figure 1.
Conversion of lignocellulosic biomasses to carbon structures.

Finally, gasification is the conversion of biomass into combustible gas by heating in air [46], pure oxygen or steam [47] at temperatures higher than 800°C with or without a catalyst [48]. Products from gasification are a mixture of carbon monoxide, carbon dioxide, methane, hydrogen and water vapour. Biochar is not the main product of gasification, but it is characterised by a simultaneous high carbon and ash content.

In summary, **Figure 1** shows the main stage of biomass conversion.

Furthermore, it is relevant to notice that leaves, stems, bark and roots are different lignin/cellulose ratio and mechanical properties. The same is true for different species.

Some of these differences could be retained into biochar and induce appreciable properties.

Also, the graphitic domains formed during pyrolytic treatment could undergo a stacking rearrangement leading to a graphitization of biochar with the increasing temperature.

3. Biochar-based reinforced plastics

Nowadays, reinforced plastic materials are one of the largest global markets in the polymer sector with an expected global revenue of up to 130 M$/year in 2024 as summarised in **Figure 2**.

Carbon-containing reinforced plastic is one of the most relevant materials with an annual production of up to 150 kton/y in 2018 [50]. As clearly reported in **Figure 3**, around 80% of the total carbon-containing reinforced plastic is represented by polymer host materials, 49% of which comes from thermoset resin and 30% from thermoplastic polymers. Among them, carbon fibre-reinforced epoxy polymers are the larger amount. This is due to their numerous applications in all-capital high-tech sectors ranging from aeronautics and aerospace industries [51] to automotive industries [52]. In this global scenario, biochar plays a minor role even if it could be used, and it is going to consolidate itself as a trustworthy commodity with its production flexibility and property tunability [53].

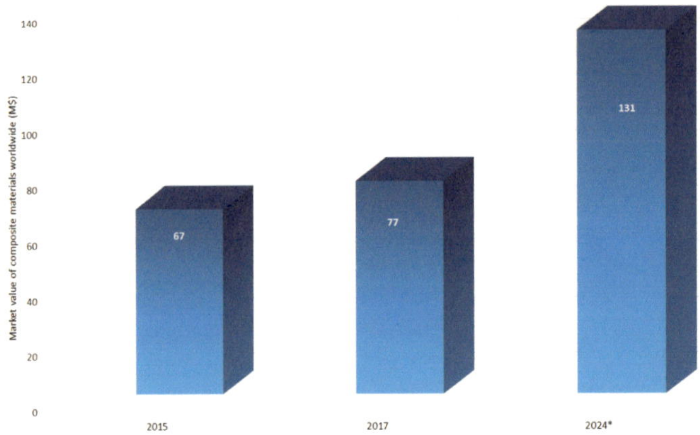

Figure 2.
Reinforced plastic worldwide revenue with a prediction for the year 2024 [49].

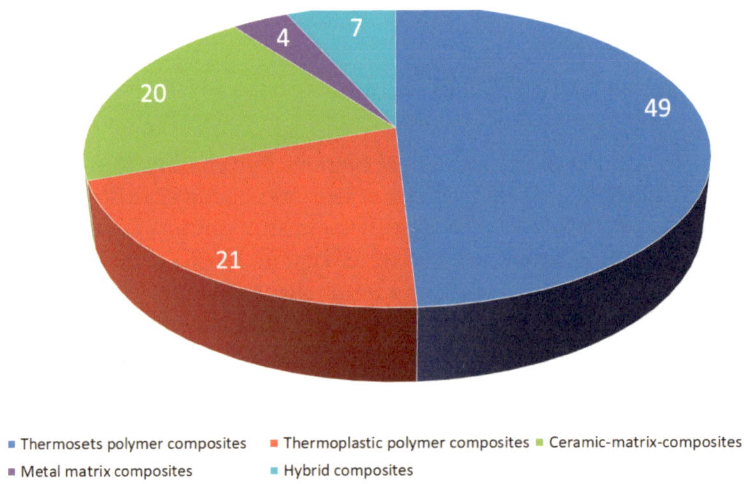

Figure 3.
World's carbon-based reinforced plastic production in 2018 [50].

Therefore, the main uses of biochar in both thermoset and thermoplastic matrices are overviewed.

Carbonaceous-reinforced thermoset resins are the most commonly used materials dispersed in plenty of different polymer hosts [54–56]. Epoxy resins are the most deeply applied and used around the world. Consequently, the replacement of carbon fibres and carbon black, carbon soot and anthracites with biochar has gained a great interest. Khan et al. [57] described the mechanical and dielectric properties of high-temperature-annealed maple-derived biochar dispersed into a two-component epoxy resin. Biochar filler was used in concentration ranging from 0.5 wt.% to 20 wt.%. The authors clearly showed the improvement of mechanical properties using filler loading of up to 4 wt.%. Regarding electric properties, Khan and co-workers found that a low loading of multiwalled CNTs induced the same effects of a 20 wt.% loading of biochar. Recently, Bartoli et al. [58] described the relationship between the biochar morphology and related composite mechanical properties using a biochar loading of 2 wt.%. The authors achieved a 40% increment of maximum elongation using a rhizomatous grass-derived biochar and Young's modulus increment using a wheat straw as a source for biochar production.

DOI: http://dx.doi.org/10.5772/intechopen.91962

The authors suggested that smooth surface could induce an improved mobility inside the epoxy matrix, while highly porous and channelled surfaces do not. Additionally, they suggested that the dispersion methodology adopted based on ultrasonication, summarised in **Figure 4**, reduced the size of biochar particles with a direct relation with the original morphology of the very same particles.

Furthermore, pyrolytic temperature is the main and critical parameter for tuning biochar properties with the goal of improving resin properties. The interactions between epoxy resin and biochar particles. Bartoli et al. [59] studied the effect of the heating rate and maximum pyrolytic temperature on biochar. Furthermore, cellulose templates could be used for the production of biochar fibres and balls using selected precursors. As a matter of fact, biochar produced from wasted cotton fibres could be recovered as carbon fibre shape showing a property enhancement of epoxy resin host matrix [60, 61], while the one produced from cellulose nanocrystals could be recovered as micrometrics ball or nanometric needles [62].

Authors showed the complex relationship of produced biochar with related containing epoxy composites. Sample prepared at different temperature and using different heating rate increment or the Young's modulus or the toughness of the reinforced plastics. Interestingly, the biochar produced at very high temperature of up to 1000°C generally induced a high increment of elongation probably due to the unpacking of the aromatic ring of epoxy host.

Similarly, Giorcelli et al. [63] proved the effectiveness of maple tree-derived biochar produced at 600°C and 1000°C, observing a drastic improvement of maximum elongation compared with neat resin.

Temperature also affected the electrical properties of biochar and biochar-containing composites. High thermal annealed biochar could represent a solid choice for the production of conductive epoxy composites. Giorcelli et al. [64] described that highly graphitic biochar induced better performances during DC electrical conductivity measurements. Temperature treatment and related graphitization processes lead to an improved ability of these materials to shield microwave radiation with similar outputs with respect to multiwalled CNTs [65] even under thin-film shape [66].

The other huge field of composite materials is represented by thermoplastic reinforced plastics. In this very same field, polyolefin represents the greater amount of worldwide production. Among them, biochar-containing polyethylene was studied by Arrigo et al. [67] using an exhausting coffee-derived biochar produced at 700°C. The authors described the rheological and thermal properties of biochar-related composites with a filler loading up to 7.5 wt. %, showing a decrement of the dynamics of polymer chains in the host matrix related to the confinement of the

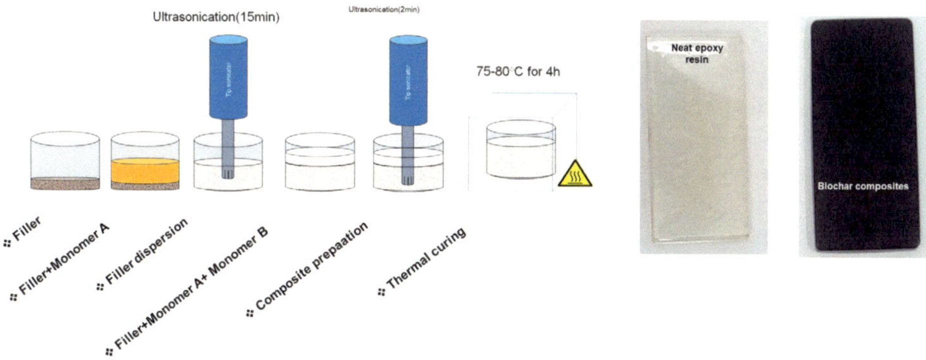

Figure 4.
Ultrasonication methodology for the effect dispersion of biochar inside the epoxy resin host.

polymer chains on the biochar surface. Additionally, the well-embedded biochar particles improved the thermo-oxidative stability of polyethylene composites produced. Zhang et al. [68] studied the temperature influence on biochar production from poplar and its use as filler for high-density polyethylene. Curiously, the microcrystalline structure of the polymer was not affected by the presence of biochar according to the thermal data collected. A different trend was reported for the mechanical properties that were appreciable different in the comparison between neat and biochar-loaded poly(ethylene) with an improvement of flexural strength and a decrement of the impact strength. Zhang et al. [69] valourized agricultural waste streams through pyrolysis, and the resulting biochar was used as filler for ultra-high-density poly(ethylene). The authors observed improvement in their mechanical properties and improvement in the flame retardancy of the high filler loading materials. Similar results were achieved by Sundarakannan et al. [70] using biochar derived from cashew nuts. Also Li et al. [71] investigated the high loads of biochar in ultra-high-molecular-weight poly(ethylene), achieving a remarkable electromagnetic interference shielding properties using an 80 wt.% of bamboo biochar pyrolysed at 1100°C. This material showed a very high conductivity of up to 107.6 S/m. Furthermore, Bajwa et al. [72] described the use of biochar for the production of a composite blend based on high-density poly(ethylene), poly(lactic acid) and wood flour with superior thermal stability.

The other largely used polyolefins is poly(propylene). Poly(propylene) is widely applied for the realisation of biochar-based composites due its workability. The main application of poly(propylene)-related composites is in the automotive sector. Tadele et al. [73] published an interesting comparative study on life cycle assessment of biochar used in automotive, showing the feasibility of its use. Das et al. [74] reached the same conclusions about the economic feasibility of the use of biochar instead of traditional carbonaceous fillers. The authors showed the appreciable cost reduction of a biochar-containing composites achieving the same properties of carbon black-based ones due to the sensible reduction of compatibilizer down to a maximum of 3 wt.%. The affordability of the cost of biochar was the core of the research proposed by Behazin et al. [75]. In this study, a pyrolysed perennial cane was used as filler of a polymer blend based on poly(propylene)/poly(octene-ethylene). The produced composite contained a filler loading ranging from 10 to 20 wt.% and showed the very strong interactions between polymer matrix and biochar particles. The most detailed and comprehensive set of studies about poly(propylene) and biochar interactions was conducted by Bhattacharyya research group and his co-workers as attested by many papers [76–78]. During this pluri-annual research, the authors investigated the kind and magnitude of interactions between filler and poly(propylene). They conclude that the addition of several types of biochars lead to a general improvement of the mechanical and thermal properties of related poly(propylene) composites together with the induction of flame retardancy properties. Additionally, Elnour et al. [79] studied the relationship between biochar properties and related poly(propylene) composites showing an increment of stiffness together with unaffected tensile strength. In the same period, Poulose et al. [80] mixed date palm-derived biochar with poly(propylene) showing the negligible effect of biochar on the storage modulus in a range of concentration of up to 15 wt.%. Poly(ethylene) and poly(propylene) are not the only polyolefins used for the production of carbon-based composites. Other largely used polyolefin matrices were poly(vinyl alcohol) and its derivatives [81, 82] and poly(acrylonitrile) [83]. Both of those two polymeric hosts are used for the realisation of piezo sensors due to their elastic properties.

Furthermore, polyesters were used for the realisation of carbon-based reinforced materials. As an example, polyamides were impregnated with biochar as described by Ogunsona et al. [84]. The authors mixed nylon 6 with the biochar produced from the pyrolysis of *Miscanthus* canes. The biochar used was produced using a process temperature ranging from 500–900°C. The different temperatures used affected the output of the composites with a beneficial effect on only the high-temperature-treated biochar and a detrimental effect on the others. In 2019, Sheng et al. [85] modified bamboo biochar through the addition of silyl groups on the particle surface for the production of poly(lactic acid) composites. Surface functionalization showing an appreciable enhancement of maximum elongation of up to 93% was compared with neat polymer matrix.

Recently, biochar was used for the realisation of biopolymer composites based on polysaccharides such as cellulose [86], starch [87] or gluten [88] under the vision of blue and green economy for a total bio- and sustainable productive line.

4. Conclusions

In this chapter, we provided overview of the most recent applications of biochar in the field of polymer composite production with a focus on more useful and unusual ones. We also described in detail the possibility of using biochars as a sound replacement for traditional fillers in both thermoset and thermoplastic composite materials. The researches herein described the feasibility of biochar used in different industrial sectors as a solid alternative to traditional and nanostructured materials. The adaptive nature of biochar presents a very strong point of advantage for spreading its use across the field of materials science.

Author details

Mattia Bartoli*, Mauro Giorcelli, Pravin Jagdale and Massimo Rovere
Department of Applied Science and Technology, Polytechnic of Turin, Torino, Italy

*Address all correspondence to: mattia.bartoli@polito.it

References

[1] Burchell TD. Carbon Materials for Advanced Technologies. Amsterdam, The Netherlands: Elsevier; 1999

[2] Holmes M. Global carbon fibre market remains on upward trend. Reinforced Plastics. 2014;**58**:38-45

[3] I.C.B. Association. What Is Carbon Black?; 2019. Availabe online: http://www.carbon-black.org/ [Accessed: 17 December 2019]

[4] Endo M. Carbon nanotube research: Past and future. Japanese Journal of Applied Physics. 2012;**51**:040001

[5] Singh V, Joung D, Zhai L, Das S, Khondaker SI, Seal S. Graphene based materials: Past, present and future. Progress in Materials Science. 2011;**56**:1178-1271

[6] Segal M. Selling graphene by the ton. Nature Nanotechnology. 2009;**4**:612

[7] Welcome To Cheap Tubes. Availabe online: https://www.cheaptubes.com/ [Accessed: 9 October 2019]

[8] Alibaba.com. Market Price for Carbon Black n550. Alibaba.com; 2019

[9] Bezama A, Agamuthu P. Addressing the Big Issues in Waste Management. London, England: SAGE Publications Sage UK; 2019

[10] Mui EL, Ko DC, McKay G. Production of active carbons from waste tyres—A review. Carbon. 2004;**42**:2789-2805

[11] Couth R, Trois C. Carbon emissions reduction strategies in Africa from improved waste management: A review. Waste Management. 2010;**30**:2336-2346

[12] Bridgwater AV. Review of fast pyrolysis of biomass and product upgrading. Biomass and Bioenergy. 2012;**38**:68-94

[13] Damartzis T, Zabaniotou A. Thermochemical conversion of biomass to second generation biofuels through integrated process design—A review. Renewable and Sustainable Energy Reviews. 2011;**15**:366-378

[14] Amen-Chen C, Pakdel H, Roy C. Production of monomeric phenols by thermochemical conversion of biomass: A review. Bioresource Technology. 2001;**79**:277-299

[15] Paz-Ferreiro J, Nieto A, Méndez A, Askeland MPJ, Gascó G. Biochar from biosolids pyrolysis: A review. International Journal of Environmental Research and Public Health. 2018;**15**:956

[16] Qian K, Kumar A, Zhang H, Bellmer D, Huhnke R. Recent advances in utilization of biochar. Renewable and Sustainable Energy Reviews. 2015;**42**:1055-1064

[17] Biochar: Prospects of Commercialization. Availabe online: https://farm-energy.extension.org/ [Accessed: 14 November 2019]

[18] Vochozka M, Maroušková A, Váchal J, Straková J. Biochar pricing hampers biochar farming. Clean Technologies and Environmental Policy. 2016;**18**:1225-1231

[19] Maroušek J. Significant breakthrough in biochar cost reduction. Clean Technologies and Environmental Policy. 2014;**16**:1821-1825

[20] Ali S, Rizwan M, Qayyum MF, Ok YS, Ibrahim M, Riaz M, et al. Biochar soil amendment on alleviation of drought and salt stress in plants: A critical review. Environmental Science and Pollution Research. 2017;**24**:12700-12712

[21] Maroušek J, Strunecký O, Stehel V. Biochar farming: Defining

economically perspective applications. Clean Technologies and Environmental Policy. 2019;**21**:1-7

[22] Maroušek J, Kolář L, Vochozka M, Stehel V, Maroušková A. Biochar reduces nitrate level in red beet. Environmental Science and Pollution Research. 2018;**25**:18200-18203

[23] Abdullah H, Wu H. Biochar as a fuel: 1. Properties and grindability of biochars produced from the pyrolysis of mallee wood under slow-heating conditions. Energy & Fuels. 2009;**23**:4174-4181

[24] Al-Wabel MI, Al-Omran A, El-Naggar AH, Nadeem M, Usman ARA. Pyrolysis temperature induced changes in characteristics and chemical composition of biochar produced from conocarpus wastes. Bioresource Technology. 2013;**131**:374-379

[25] Chen W-H, Peng J, Bi XT. A state-of-the-art review of biomass torrefaction, densification and applications. Renewable and Sustainable Energy Reviews. 2015;**44**:847-866

[26] Gronnow MJ, Budarin VL, Mašek O, Crombie KN, Brownsort PA, Shuttleworth PS, et al. Torrefaction/biochar production by microwave and conventional slow pyrolysis—Comparison of energy properties. GCB Bioenergy. 2013;**5**:144-152

[27] Bach Q-V, Chen W-H, Chu Y-S, Skreiberg Ø. Predictions of biochar yield and elemental composition during torrefaction of forest residues. Bioresource Technology. 2016;**215**:239-246

[28] Bartoli M, Rosi L, Giovannelli A, Frediani P, Frediani M. Characterization of bio-oil and bio-char produced by low-temperature microwave-assisted pyrolysis of olive pruning residue using various absorbers. Waste Management & Research. 2020;**38**(2):213-225

[29] Bartoli M, Rosi L, Giovannelli A, Frediani P, Frediani M. Bio-oil from residues of short rotation coppice of poplar using a microwave assisted pyrolysis. Journal of Analytical and Applied Pyrolysis. 2016;**119**:224-232

[30] Bartoli M, Rosi L, Giovannelli A, Frediani P, Frediani M. Pyrolysis of α-cellulose in a microwave multimode batch reactor. Journal of Analytical and Applied Pyrolysis. 2016;**120**:284-296

[31] Bartoli M, Rosi L, Giovannelli A, Frediani P, Frediani M. Production of bio-oils and bio-char from Arundo donax through microwave assisted pyrolysis in a multimode batch reactor. Journal of Analytical and Applied Pyrolysis. 2016;**122**:479-489

[32] Wampler TP. Applied Pyrolysis Handbook. Boca Raton, FL, USA: CRC Press; 2006

[33] Bridgwater A, Peacocke G. Fast pyrolysis processes for biomass. Renewable and Sustainable Energy Reviews. 2000;**4**:1-73

[34] Ragucci R, Giudicianni P, Cavaliere A. Cellulose slow pyrolysis products in a pressurized steam flow reactor. Fuel. 2013;**107**:122-130

[35] Tsai W, Lee M, Chang Y. Fast pyrolysis of rice straw, sugarcane bagasse and coconut shell in an induction-heating reactor. Journal of Analytical and Applied Pyrolysis. 2006;**76**:230-237

[36] Karaduman A, Şimşek EH, Çiçek B, Bilgesü AY. Flash pyrolysis of polystyrene wastes in a free-fall reactor under vacuum. Journal of Analytical and Applied Pyrolysis. 2001;**60**:179-186

[37] Li D, Briens C, Berruti F. Improved lignin pyrolysis for phenolics production in a bubbling bed reactor—Effect of bed materials. Bioresource Technology. 2015;**189**:7-14

[38] Horne PA, Williams PT. Influence of temperature on the products from the flash pyrolysis of biomass. Fuel. 1996;**75**:1051-1059

[39] Zhang X, Yang W, Dong C. Levoglucosan formation mechanisms during cellulose pyrolysis. Journal of Analytical and Applied Pyrolysis. 2013;**104**:19-27

[40] Kotake T, Kawamoto H, Saka S. Mechanisms for the formation of monomers and oligomers during the pyrolysis of a softwood lignin. Journal of Analytical and Applied Pyrolysis. 2014;**105**:309-316

[41] Behrendt F, Neubauer Y, Oevermann M, Wilmes B, Zobel N. Direct liquefaction of biomass. Chemical Engineering & Technology. 2008;**31**:667-677

[42] Akhtar J, Amin NAS. A review on process conditions for optimum bio-oil yield in hydrothermal liquefaction of biomass. Renewable and Sustainable Energy Reviews. 2011;**15**:1615-1624

[43] Liu Z, Zhang F-S. Effects of various solvents on the liquefaction of biomass to produce fuels and chemical feedstocks. Energy Conversion and Management. 2008;**49**:3498-3504

[44] Zou S, Wu Y, Yang M, Li C, Tong J. Thermochemical catalytic liquefaction of the marine microalgae *Dunaliella tertiolecta* and characterization of bio-oils. Energy & Fuels. 2009;**23**:3753-3758

[45] Peterson AA, Vogel F, Lachance RP, Fröling M, Antal MJ Jr, Tester JW. Thermochemical biofuel production in hydrothermal media: A review of sub- and supercritical water technologies. Energy & Environmental Science. 2008;**1**:32-65

[46] Guizani C, Sanz FE, Salvador S. Influence of temperature and particle size on the single and mixed atmosphere gasification of biomass char with H_2O and CO_2. Fuel Processing Technology. 2015;**134**:175-188

[47] Barisano D, Canneto G, Nanna F, Alvino E, Pinto G, Villone A, et al. Steam/oxygen biomass gasification at pilot scale in an internally circulating bubbling fluidized bed reactor. Fuel Processing Technology. 2016;**141**:74-81

[48] Cheah S, Jablonski WS, Olstad JL, Carpenter DL, Barthelemy KD, Robichaud DJ, et al. Effects of thermal pretreatment and catalyst on biomass gasification efficiency and syngas composition. Green Chemistry. 2016

[49] Grand View Research. Market Value of Composite Materials Worldwide from 2015 to 2024 (in Billion U.S. Dollars). Grand View Research, statista.com; 2019

[50] Sauer M, Kühnel M, Witten E. Composites Market Report 2018—Market Developments, Trends, Outlook and Challenges. Germany: AVK-TV Industrievereinigung Verstrkte Kunststoffe Carbon Composite; 2018

[51] Chand S. Review carbon fibers for composites. Journal of Materials Science. 2000;**35**:1303-1313

[52] Adam H. Carbon fibre in automotive applications. Materials & Design. 1997;**18**:349-355

[53] Downie A, Crosky A, Munroe P. Physical properties of biochar. In: Biochar for Environmental Management: Science and Technology. London, UK: MPG Book; 2009. pp. 13-32

[54] Bartoli M, Rosi L, Frediani M. Chapter 22—Synthesis and applications of unsaturated polyester composites. In: Thomas S, Hosur M, Chirayil CJ, editors. Unsaturated Polyester Resins. Amsterdam, The Netherlands: Elsevier; 2019. pp. 579-598

[55] Rana S, Alagirusamy R, Joshi M. A review on carbon epoxy nanocomposites. Journal of Reinforced Plastics and Composites. 2009;**28**:461-487

[56] Kim S, Pechar TW, Marand E. Poly (imide siloxane) and carbon nanotube mixed matrix membranes for gas separation. Desalination. 2006;**192**: 330-339

[57] Khan A, Savi P, Quaranta S, Rovere M, Giorcelli M, Tagliaferro A, et al. Low-cost carbon fillers to improve mechanical properties and conductivity of epoxy composites. Polymers. 2017;**9**:642

[58] Bartoli M, Giorcelli M, Rosso C, Rovere M, Jagdale P, Tagliaferro A. Influence of commercial biochar fillers on brittleness/ductility of epoxy resin composites. Applied Sciences. 2019;**9**:3109

[59] Bartoli M, Nasir MA, Jagdale P, Passaglia E, Spiniello R, Rosso C, et al. Influence of pyrolytic thermal history on olive pruning biochar and related epoxy composites mechanical properties. Journal of Composite Materials. 2019:11

[60] Jagdale P, Koumoulos EP, Cannavaro I, Khan A, Castellino M, Dragatogiannis DA, et al. Towards green carbon fibre manufacturing from waste cotton: A microstructural and physical property investigation. Manufacturing Review. 2017;**4**:10

[61] Khan A, Jagdale P, Castellino M, Rovere M, Jehangir Q, Mandracci P, et al. Innovative functionalized carbon fibers from waste: How to enhance polymer composites properties. Composites Part B: Engineering. 2018;**139**:31-39

[62] Bartoli M, Giorcelli M, Jagdale P, Rovere M, Tagliaferro A, Chae M, et al. Shape tunability of carbonized cellulose nanocrystals. SN Applied Sciences. 2019;**1**:1661-1676

[63] Giorcelli M, Khan A, Pugno NM, Rosso C, Tagliaferro A. Biochar as a cheap and environmental friendly filler able to improve polymer mechanical properties. Biomass and Bioenergy. 2019;**120**:219-223

[64] Giorcelli M, Savi P, Khan A, Tagliaferro A. Analysis of biochar with different pyrolysis temperatures used as filler in epoxy resin composites. Biomass and Bioenergy. 2019;**122**:466-471

[65] Giorcelli M, Khan A, Tagliaferro A, Savi P, Berruti F. Microwave characterization of polymer composite based on biochar: A comparison of composite behaviour for biochar and MWCNTs. In: 2016 IEEE International Nanoelectronics Conference (INEC). Chengdu, China: IEEE; 9-11 May 2016. pp. 1-2

[66] Quaranta S, Savi P, Giorcelli M, Khan AA, Tagliaferro A, Jia CQ. Biochar-polymer composites and thin films: Characterizations and applications. In: 2016 IEEE 2nd International Forum on Research and Technologies for Society and Industry Leveraging a Better Tomorrow (RTSI); 2016. pp. 1-4

[67] Arrigo R, Jagdale P, Bartoli M, Tagliaferro A, Malucelli G. Structure–property relationships in polyethylene-based composites filled with biochar derived from waste coffee grounds. Polymers. 2019;**11**:1336

[68] Zhang Q, Khan MU, Lin X, Cai H, Lei H. Temperature varied biochar as a reinforcing filler for high-density polyethylene composites. Composites Part B: Engineering. 2019;**175**:107151

[69] Zhang Q, Zhang D, Xu H, Lu W, Ren X, Cai H, et al. Biochar filled high-density polyethylene composites with excellent properties: Towards maximizing the utilization of agricultural wastes. Industrial Crops and Products. 2020;**146**:112185

[70] Sundarakannan R, Arumugaprabu V, Manikandan V, Vigneshwaran S. Mechanical property analysis of biochar derived from cashew nut shell waste reinforced polymer matrix. Materials Research Express. 2020;**6**:125349

[71] Li S, Huang A, Chen Y-J, Li D, Turng L-S. Highly filled biochar/ultra-high molecular weight polyethylene/linear low density polyethylene composites for high-performance electromagnetic interference shielding. Composites Part B: Engineering. 2018;**153**:277-284

[72] Bajwa DS, Adhikari S, Shojaeiarani J, Bajwa SG, Pandey P, Shanmugam SR. Characterization of bio-carbon and ligno-cellulosic fiber reinforced bio-composites with compatibilizer. Construction and Building Materials. 2019;**204**:193-202

[73] Tadele D, Roy P, Defersha F, Misra M, Mohanty AK. A comparative life-cycle assessment of talc-and biochar-reinforced composites for lightweight automotive parts. Clean Technologies and Environmental Policy. 2020:1-11

[74] Das O, Bhattacharyya D, Sarmah AK. Sustainable eco–composites obtained from waste derived biochar: A consideration in performance properties, production costs, and environmental impact. Journal of Cleaner Production. 2016;**129**:159-168

[75] Behazin E, Misra M, Mohanty AK. Sustainable biocarbon from pyrolyzed perennial grasses and their effects on impact modified polypropylene biocomposites. Composites Part B: Engineering. 2017;**118**:116-124

[76] Das O, Kim NK, Kalamkarov AL, Sarmah AK, Bhattacharyya D. Biochar to the rescue: Balancing the fire performance and mechanical properties of polypropylene composites. Polymer Degradation and Stability. 2017;**144**:485-496

[77] Das O, Sarmah AK, Bhattacharyya D. Biocomposites from waste derived biochars: Mechanical, thermal, chemical, and morphological properties. Waste Management. 2016;**49**:560-570

[78] Das O, Bhattacharyya D, Hui D, Lau K-T. Mechanical and flammability characterisations of biochar/polypropylene biocomposites. Composites Part B: Engineering. 2016;**106**:120-128

[79] Elnour AY, Alghyamah AA, Shaikh HM, Poulose AM, Al-Zahrani SM, Anis A, et al. Effect of pyrolysis temperature on biochar microstructural evolution, physicochemical characteristics, and its influence on biochar/polypropylene composites. Applied Sciences. 2019;**9**:1149

[80] Poulose AM, Elnour AY, Anis A, Shaikh H, Al-Zahrani SM, George J, et al. Date palm biochar-polymer composites: An investigation of electrical, mechanical, thermal and rheological characteristics. Science of the Total Environment. 2018;**619-620**:311-318

[81] Nan N, DeVallance DB, Xie X, Wang J. The effect of bio-carbon addition on the electrical, mechanical, and thermal properties of polyvinyl alcohol/biochar composites. Journal of Composite Materials. 2016;**50**:1161-1168

[82] Nan N, DeVallance DB. Development of poly (vinyl alcohol)/wood-derived biochar composites for use in pressure sensor applications. Journal of Materials Science. 2017;**52**:8247-8257

[83] Nan W, Zhao Y, Ding Y, Shende AR, Fong H, Shende RV. Mechanically flexible electrospun carbon nanofiber mats derived from biochar and

polyacrylonitrile. Materials Letters. 2017;**205**:206-210

[84] Ogunsona EO, Misra M, Mohanty AK. Impact of interfacial adhesion on the microstructure and property variations of biocarbons reinforced nylon 6 biocomposites. Composites Part A: Applied Science and Manufacturing. 2017;**98**:32-44

[85] Sheng K, Zhang S, Qian S, Fontanillo Lopez CA. High-toughness PLA/bamboo cellulose nanowhiskers bionanocomposite strengthened with silylated ultrafine bamboo-char. Composites Part B: Engineering. 2019;**165**:174-182

[86] Zhang Q, Lei H, Cai H, Han X, Lin X, Qian M, et al. Improvement on the properties of microcrystalline cellulose/polylactic acid composites by using activated biochar. Journal of Cleaner Production. 2020;**252**:119898

[87] George J, Azad LB, Poulose AM, An Y, Sarmah AK. Nano-mechanical behaviour of biochar-starch polymer composite: Investigation through advanced dynamic atomic force microscopy. Composites Part A: Applied Science and Manufacturing. 2019;**124**:105486

[88] Das O, Hedenqvist MS, Johansson E, Olsson RT, Loho TA, Capezza AJ, et al. An all-gluten biocomposite: Comparisons with carbon black and pine char composites. Composites Part A: Applied Science and Manufacturing. 2019;**120**:42-48

Chapter 5

POSS Fillers as a Factor Influencing on Viscoelastic Properties, Crystallization, and Thermo-Oxidative Degradation of Poly(Lactic Acid)-Epoxidized Natural Rubber PLA/ENR Blend

Magdalena Lipińska, Klaudia Toczek and Magdalena Stefaniak

Abstract

Polyhedral oligomeric silsesquioxanes (POSS) can be considered as the smallest silica particles that contain an inorganic cage-like silicon-oxygen structure surrounded by organic substituents. POSS are known to be reinforcing fillers able to enhance the fire retardancy and to increase the thermo-oxidative stability. The application of POSS with various functional groups and their effect on the properties of poly(lactic acid) (PLA) is analyzed based on the literature review. The influence of POSS with hydroxyl and glycidyl groups on the viscoelastic properties, crystallization, and thermo-oxidative degradation of PLA containing various contents of natural rubber ENR is discussed. The application of POSS with hydroxyl and glycidyl groups as an additive enhancing compatibility of PLA toughened by epoxidized rubber (ENR) is described. It is reported that POSS molecules with hydroxyl groups can enhance the thermo-oxidative stability of PLA melt.

Keywords: POSS, poly(lactic acid), epoxidized natural rubber ENR, viscoelastic properties, thermo-oxidative stability

1. Introduction

Nowadays polyhedral oligomeric silsesquioxanes (POSS) particles has received growing attentions as a new type of nanofillers used to prepare nanocomposites with enhanced thermal or mechanical properties [1, 2].

POSS has hybrid organic-inorganic structure, and it can be considered as the smallest functionalized silica particle containing a cage-like silicon-oxygen inorganic core surrounded by organic groups. Typical POSS structure has the formula $(RSiO_{1.5})_n$, where n = 6, 8, 10, and R are organic substituents at the corners of the inorganic cage [3]. The most common POSS structure has n = 8. Many research reports have been devoted to the various aspects of the application of POSS.

IntechOpen

The inorganic core of POSS may provide molecular reinforcement, while the organic functional groups may increase the compatibility with polymeric matrix or be incorporated into polymeric network via co- and polymerization reactions [4].

POSS nanoparticles due to their flexible chemical and physical hybrid properties can be applied as functional additives able to reinforce, to toughen, and to stabilize various polymeric matrices, among them are polypropylene [5], polystyrene [6], polyamide [7], polyurethanes [8], and polysiloxane elastomer [9]. Herein the incorporation of POSS particles, which provide remarkable reinforcement at low filler loading, is advantageous and beneficial. Additionally POSS can be grafted to the polymer chains via chemical reactions of functional organic groups leading to the formation of chemical bonds [10]. POSS particles offer the possibility for the participation in curing by a reaction between POSS functional groups and the end or functional groups of polymer chains. The radicals generated during peroxide curing were used to initiate the grafting reactions between methacrylisobutyl-POSS and octavinyl-POSS and ethylene-propylene elastomer/hydrogenated acrylonitrile rubber EPM/HNBR blends [11]. The incorporation of octavinyl-POSS particles, which were able to form crosslinks between both rubbers, enhanced the compatibility of EPM/HNBR blend. Glycidyl polyhedral oligomeric silsesquioxane POSS was utilized as a crosslinking additive to prepare an organic-inorganic hybrid material based on the carboxyl-terminated poly(acrylonitrile-*co*-butadiene) [12]. In this work the trisilanolisobutyl-POSS and glycidyl-POSS were incorporated into the epoxidized natural rubber/poly(lactic acid) blend to enhance the mechanical, thermal, and oxidative properties. The influence of both POSS molecules on the thermo-oxidative degradation of polylactide is analyzed.

2. POSS structures as reinforcing additives to poly(lactic acid)

Thermoplastic material such as poly(lactic acid) usually is processed into fiber and films. Likewise many other aliphatic polyesters are an emerging polymer from renewable sources. In general, PLA is biodegradable and nontoxic for human body and thus is used for implantable composites, controlled drug release, food-packaging materials, and systems for sustained release of pesticides and fertilizers [13]. The high strength, modulus, and biodegradation make PLA a promising material for the production of packaging materials, although the brittleness restricts its large-scale application as an environmental friendly plastic product. Due to the chirality of lactic acid, various forms of polylactide exist as poly(L-lactide) PLLA or PLA, poly(D-lactide) PDLA, poly(D,L-lactide) PDLA, and poly(L-lactide-*co*-D,L-lactide) PLDLLA, which vary in properties [14–16]. PLLA is the product of L,L-lactide polymerization. It has a glass transition temperature T_g in a range of 50–65°C, a melting temperature of about 173–178°C, a crystallinity of around 37%, and a tensile modulus between 2.7 and 16 GPa [17, 18]. The blending of PLLA with PDLA allows the increase of the melting temperature and the temperature stability of the blend; additionally PDLA acts as a nucleating agent increasing the crystallization rate [14, 19, 20]. Biodegradation of the polylactide occurs due to a simple hydrolysis as the main degradation mechanism; thus it depends on the chirality of various PLA forms, and it is slower for PDLA because of the higher crystallinity [21].

Various methods were applied to obtain the POSS/poly(lactic acid) composites, among them are the solution casting method [22], the melt blending method [23–27], and injection molding [28].

These reports confirmed that the incorporation of various POSS particles had a reinforcing effect on the polylactide PLA properties. The crucial significance of the chemistry, reactivity, or non-reactivity of the functional POSS groups was observed. These groups can control and enhance the compatibility between POSS and poly(lactic acid) leading to the better mechanical performance.

Octaisobutyl-POSS, aminopropylisobutyl-POSS, trisilanolisobutyl-POSS, and glycidylisobutyl-POSS/poly(ethylene-glycol) plasticized PLA composites, having various POSS content (1, 3, 7, and 10 wt%), were prepared [28]. The presence of different functional groups in POSS structure affected the dispersion and hence the mechanical properties of composites. Kodal et al. [28] reported the enhancement of the impact strength. Toughness of the plasticized poly(lactic acid) was significantly affected by the incorporation of POSS particles leading to the increase of the elongation at break. Moreover, it was revealed that regardless of POSS type, the melt viscosity of the composites decreased, in particular for the glycidylisobutyl-POSS. The presence of functional groups and its polarity was a crucial factor influencing the compatibility between POSS and polymer leading to the changes in mechanical performance. With the increasing POSS amount (from 1 to 10 wt%) incorporated into the polymer matrix, the tensile modulus of POSS/PEG/PLA composites decreased due to the worse dispersion of POSS particles at higher loading. Opposite the elongation at break, the energy at break and Izod impact strength meaningfully increased, indicating better toughness of the material after the incorporation of POSS particles.

A poly(ethylene glycol)-POSS containing long and flexible PEG segments attached to the inorganic core were prepared and applied to modify the brittleness of PLA [29, 30]. The toughening effect of PEG-POSS was observed [30]. The stress-strain curves of neat PLLA showed the typical brittle polymer tensile behavior, whereas for the PLLA containing 10 wt% of PEG-POSS, the shape of the curve was changed to atypical for ductile polymers. The tensile strength of the PEG-POSS/PLLA was reduced from 41.4 (neat PLLA) to 21.6 MPa; simultaneously the elongation of break increased from 8.7 (neat PLLA) to 21.6% [30]. The effect was attributed to the higher chain mobility of the PEG-POSS/PLLA composites as a PEG-POSS particles acted as an efficient plasticizing agent.

Zhang et al. [31] investigated the mechanical properties of octa(3-chloropropylsilsesquioxane)/PLLA hybrid films containing 3, 5, 10 and 15 wt% of incorporated POSS. The octa(3-chloropropylsilsesquioxane) was dispersed at molecular level at 3 wt%; higher loading led to aggregation and the crystallization of POSS particles. The tensile behavior of the PLLA material changed. More ductile behavior instead of the brittle behavior was observed. Elongation of break increased significantly from 13.5 (neat PLLA) to 91.0% (PLLA), containing 15 wt% of POSS confirming that the PLLA matrix was more plasticized [31]. Furthermore the incorporation of POSS caused the shifting of the glass transition temperature T_g and the melting temperatures of the PLLA matrix to lower values of temperature.

PLA composites having various octaisobutyl-POSS content (1, 3, 7, and 10 wt%) were obtained using a twin-screw micro-compounder [24]. It was showed that the isobutyl groups present in POSS increased compatibility with PLA backbone groups resulting in higher homogeneity and the miscibility between both components. Additionally, Yilmaz et al. [24] demonstrated that the incorporation of the octaisobutyl-POSS affected two components of total fracture work: essential work of fracture (the work spent in the inner fracture process zone) and nonessential work of fracture (the work spent in the plastic deformation zone). The unstable and rapid crack growth occurred suddenly after the maximum load in the neat PLA. The presence of octaisobutyl-POSS in PLA matrix raised the ductility of the composite, resulting in a stable crack growth for all POSS loadings. The greater energy absorption, the enhancement in toughness, and the increased extension to break without a significant loss of Young's modulus were attributed to the plasticization of PLA caused by the presence of octaisobutyl-POSS. The chain mobility of PLA was facilitated, as confirmed by the reduction in the glass transition temperature T_g of composites [24].

It was found also that the addition of octaisobutyl-POSS (5 wt%) led to the enhancement of the storage modulus (E′) below the glass transition temperature T_g; as compared with neat PLLA, the values of E′ increased from 1875 (neat PLLA) to 2485 MPa POSS/PLLA composite at 20°C [22]. The segmental motions of polymer chains in POSS/PLLA composite were not restricted, and no significant shift of the glass transition temperature T_g was observed [22].

Turan et al. [23] observed the plasticizing effect of POSS containing seven isobutyl groups and one functional aminopropyl group on PLA. Aminopropylisobutyl-POSS (1, 3, 10 wt%) reduced the melt viscosity of PLA composites. The power consumption during melt processing lowered as the processing was facilitated. This could be the additional advantage of POSS application. Tensile strength studies showed that the lower 1, 3 wt% content of aminopropylisobutyl-POSS enhanced Young's modulus. As the POSS content increased to 3 wt%, the modulus increased to 40% [23]. On the other hand, the incorporation of aminopropylisobutyl-POSS reduced the strain at break and yield point of PLA composites [23].

The formation of covalent bonds and physical interactions via hydrogen bonds between POSS and PLA chains usually lead to the enhancement of dynamic mechanical properties of POSS/PLA composites. This effect was observed for aminopropylisobutyl-POSS/PLA composites [23] up to 3 wt% of POSS content due to the physical (H-bonding) interactions of PLA chains with—NH_2 functional groups [23]. Higher content (10 wt%) of POSS particles caused the deterioration of dynamic mechanical properties resulting from the agglomeration of the filler.

In order to improve the compatibility and lead to better mechanical properties, the hybrid POSS-PLLA nanoparticles with different tail lengths were incorporated to neat PLLA. PLLA/POSS nanocomposites, in which the PLLA chains were grafted onto the POSS molecules, were prepared via microwave-assisted ring-opening polymerization [32]. The interfacial compatibility was strongly affected by the length of grafted PLLA chain and the amount of PLLA-POSS added. The addition of 6 wt% of PLLA-POSS with the longest grafted PLLA tail leading to effective nucleation activity and the improvement of Young's modulus by 57% and the tensile strength by 26.5%. Other authors [33] observed simultaneous enhancement in toughness and strength for polylactide containing octa-armed polyhedral oligomeric silsesquioxanes POSS-$(PLLA)_8$ star hybrid particles.

The ability to increase the compatibility between components of blends is another important factor that the application of POSS offers.

Blends of poly(lactic acid) PLA, triallyl isocyanurate, and octavinyl-POSS were prepared and cured by γ-irradiation [27]. The presence of the unsaturated functional groups in POSS structure provided the possibility of the grafting of POSS moieties to the PLA chains. The crosslinking occurred together with the degradation processes. The chemical bonds were formed between the double bonds of POSS particles and the polymer radical generated during γ-irradiation exposure. Due to the higher crosslink density of octavinyl-POSS/PLA composites the tensile strength, E-modulus, and the heat deflection temperature increased. The glass transition temperatures of POSS/PLA composites were shifted to higher temperature confirming the formation of more cross-linked network in the presence of octavinyl-POSS.

Other authors [25] incorporated 2 and 5 wt% of octavinyl-POSS and epoxycyclohexyl-POSS to blend prepared by mixing of poly(lactic acid) and poly(butylene succinate-co-adipate). PLA/PBSA blend was prepared using a fixed weight ratio of 70:30 of both components [25]. The opposite effects of both POSS on the viscoelastic properties, the storage modulus and complex viscosity, were observed. Octavinyl-POSS acted as a plasticizing additive, reducing the melt elasticity and decreasing the values of the storage modulus and complex viscosity. The epoxycyclohexyl group of POSS reacted with the functional groups present at the end of

PLA/PBSA chains leading to the enhancement of the storage modulus and higher compatibility of the blend.

Octaisobutyl-POSS and trans-cyclohexanediolisobutyl POSS containing functional hydroxyl groups potentially that are able to interact with both polymer matrices were added at 2 wt% loading to poly(lactide) and poly(ε-caprolactone) blend by melt blending method [34]. Moreover, to enhance the POSS adhesion to the components of the blends, authors [34] grafted the PCL-*b*-PLLA block copolymer onto the aminopropyl heptaisobutyl-POSS molecules via ring-opening polymerization of ε-caprolactone and L-lactide. The functionalized POSS molecules were found to be compatibilizing agent able to stabilize the morphology of PLA/PCL blends [34].

The compatibilizing and stabilizing effect of various particles reported in literature can be attributed to the decrease of the interfacial energy; the localization of the solid particles at the interphase between two polymers, acting as a solid barrier preventing the coalescence; the strong interactions with polymeric chains bringing about the steric hindrance; and the change in the viscosity ratio resulting from the uneven distribution of filler particles in both phases. Monticelli et al. [34] reported that the octaisobutyl-POSS was preferentially located in PCL phase resulting in the increase of polymer viscosity. Higher viscosity reduced the coalescence leading to the reduction of the PCL droplets size. Trans-cyclohexanediolisobutyl POSS because of its higher functionality and affinity to both phases was able to locate at the interphase leading to stronger interfacial adhesion of both components in PLA/PCL composite.

Han et al. [35] reported the compatibilizing effect of Janus POSS star hybrid, which contained an inorganic POSS core and three poly(ε-caprolactone) arms and five poly(L-lactide) arms, on the PLLA/PCL (w/w = 70/30) blends. Due to the reduction in interfacial tension, the dispersed PCL domain size decreased from 1 to 0.5 μm. Better compatibility caused by POSS star particles led to the increase of the Izod impact strength and tensile mechanical properties.

The reduced interfacial tension in poly(lactic acid)/starch/poly(ε-caprolactone) blends after the incorporation of the trifluoropropyl-POSS particles (1, 3, 5 wt%) was also the main factor leading to the enhanced compatibility of the system [26].

Summarizing the application of various POSS molecules, it is worth to underline that the reinforcing effect of POSS molecules is strongly dependent on the chemistry of the organic groups present in the filler structure. The compatibility between the POSS and PLA can be enhanced by the presence in POSS backbone aliphatic groups with higher affinity to polymer chains. These groups can facilitate mixing between polymer and filler during processing reducing, uneven dispersion of filler in matrix and leading to higher homogeneity. Thus the better dispersion improves mechanical properties. On the other hand, more polar groups, in particular glycidyl, epoxy, or hydroxyl groups, can form physical interactions or even covalent bonds with the end groups of PLA, causing the improvement of dynamic mechanical properties.

Herein we decided to apply POSS molecules containing isooctyl, hydroxyl, and glycidyl functionalities as an additive to epoxidized natural rubber PLA blends.

3. Influence of POSS on crystallinity, thermal stability, and degradation of poly(lactic acid)

PLLA is a semicrystalline polymer; thus its crystallization behavior and degree of crystallinity are important factors playing a vital role in various properties, e.g., mechanical, thermal, and degradation properties. A large number of studies are reported in the literature related to the crystallization of poly(lactic acid) [36–39]. The crystallization kinetics [40], the influence of the molecular weight and undercooling on morphology and crystal grow [41], and the isothermal melting mechanism [42] were investigated. PLA crystallization, even at high L-LA content, is typically too slow; thus, to increase the crystallinity, three main strategies can be considered: the incorporation of a nucleating agent, the addition of a plasticizer, and the adjustment of the molding conditions, in particular molding temperature and cooling time. Talc is usually used as a nucleating agent that lowers the surface-free energy barrier toward nucleation, increasing the crystallization rate [39].

Some authors reported a nucleating influence of aminopropylisobutyl-POSS particles on the crystallization [23]. The effect was confirmed by the enhanced degree of crystallinity and lower cold crystallization temperatures for PLA composites containing 3 and 10 wt% of POSS [23].

Wang et al. [25] showed that octavinyl-POSS, and epoxycyclohexyl-POSS can act as nucleating agents, enhancing the crystallization ability of poly(lactic acid) in PLA/PBSA blends. Additionally both incorporated POSS increased the decomposition temperature retarding the thermal degradation of PLA/PBSA blends.

Qiu et al. [22] investigated poly(L-lactide)/octaisobutyl-POSS nanocomposites prepared via solution casting method. The crystallization studies revealed that 5 wt% of octaisobutyl-POSS increased the degree of crystallinity from around 14.8% for neat PLLA to around 43.2% for the PLLA/octaisobutyl-POSS composites. Octaisobutyl-POSS reduced the half time of crystallization $t_{0.5}$ defined as the time required to achieve 50% of the final crystallinity for the sample. The nucleation density was enhanced in the PLLA/octaisobutyl-POSS composites compared with that of neat PLLA. Octaisobutyl-POSS increased the values of weight loss with prolonged exposure time, indicating the accelerating effect of the POSS particles on the hydrolytic degradation of PLLA in composite.

Other authors [28] also reported the influence of POSS with octaisobutyl, amine, hydroxyl, and epoxide groups on the crystallization behavior of PEG-plasticized PLA. All POSS types reduced the cold crystallization temperature T_{cc}, leading to much faster nucleation rate of POSS-filled PLA and PEG-plasticized PLA. Furthermore the POSS addition shifted the glass transition temperature of PLA phase to lower value of temperature due to the plasticization of PLA matrix. The strongest plasticizing effect was observed for octaisobutyl-POSS. All utilized POSS particles enhanced the thermal stability acting as a physical barrier which reduced the heat flow to the matrix and increased the decomposition temperatures at 5%, 10%, and maximum weight loss.

Yu and Qiu [43] studied the crystallization behavior, spherulitic morphology, and crystal structure of poly($_L$-lactide) mixed via solution casting method with two different weight ratios (0.5, 1 wt%) of octavinyl-POSS. Both the nonisothermal melt and cold crystallization of PLLA were enhanced by the presence of octavinyl-POSS presence, to the higher degree of crystallinity χ_C = 44.4% for 1 wt% of octavinyl-POSS compared to χ_C = 8.0% for neat PLLA. The overall crystallization rates during isothermal crystallization were faster and improved with increasing octavinyl-POSS content. The time to achieve the 50% progress of crystallization shortened, and the calculated values of $t_{0.5}$ were lower for octavinyl-POSS/PLLA composites confirming that the crystallization of PLLA was accelerated in the presence of POSS particles. No significant changes in the crystallization mechanism were found, although the number and the size of developed spherulites changed indicating the nucleating effect caused by the POSS particles. The nucleation density of PLLA spherulites increased significantly in the octavinyl-POSS/PLLA in respect to the neat PLLA. The neat PLLA and octavinyl-POSS/PLLA composites showed a similar degradation profile. The degradation temperatures at 5% of weight loss and the decomposition temperatures were reduced slightly after the incorporation of octavinyl-POSS particles [43].

Similar effect on crystallinity was reported for the PLLA composites containing various octa(3-chloropropylsilsesquioxane) content [31]. Studies indicated that octa(3-chloropropyl)-POSS can be an effective heterogeneous nucleating agent promoting the crystallization of PLLA and increasing the degree of crystallinity.

The low concentration of fluorinated POSS particles (0.5 wt%) was found to be a suitable nucleating agent, increasing the crystallinity of PLA in PLA/starch/PCL blend [26]. At a low concentration POSS acted as a lubricating agent that lowered the viscosity of polymer melt which had the positive effect on the nucleation and growth stages. At higher POSS concentration the increased viscosity of the system had an adverse influence on the growth of spherulites [26].

As we previously described, silsesquioxanes with reactive functionalities are suitable for polymerization, surface bonding, or grafting. The POSS molecules containing hydroxyl groups could be adopted as initiators of the ring-opening polymerization ROP during the synthesis of PLA from lactide. Organic-inorganic hybrids with various arms' length of poly(lactic acid) tethered with POSS synthetized via ring-opening polymerization of L-lactide were found to be nucleating additives increasing the degree of crystallinity of PLA [32, 33].

Similar to other aliphatic polyesters, PLA is able to degrade during processing because of the action of the external factors. Polymer degradation can occur due to the influence of the temperature (thermal degradation), the effect of the presence of oxygen, and the elevated temperature (thermo-oxidative degradation), as a result of the mechanical stress (thermomechanical degradation). During the processing at the elevated temperature (usually in temperature range of 180–200°C), the thermo-oxidative degradation can lead to the random chain scission and the formation of degradation products [44].

POSS molecules were found to be the additives which according to the chemical structure can both accelerate or postpone the thermal degradation of PLA composites. The influence of the octaisobutyl-POSS and the accelerating effect of POSS incorporation on the hydrolytic degradation of POSS/PLLA nanocomposite with respect to neat PLLA were reported [22]. Octavinyl-POSS caused the opposite effect leading to the higher thermal stability and the delay in oxidation of PLA [45].

4. Materials and characterization

L-polylactide PLA, CAS number 331335-50-1 product of Simagchem Corporation (China) was applied as a dominant component of the poly(lactic acid)/epoxidized natural rubber blends, the formulations of the blends were as follows:

100 g PLA—ENR 75 g, further denoted as 75 ENR/PLA blend
100 g PLA—ENR 50 g, further denoted as 50 ENR/PLA blend
100 g PLA—ENR 25 g, further denoted as 25 ENR/PLA blend

In this study the epoxidized natural rubber ENR50 (Epoxyprene 50, Muang Mai Guthrie Company) was used.

A liquid glycidyl-POSS cage mixture EP0409 (Hybrid, USA), further denoted as Gly-POSS, with an inorganic silsesquioxane core and eight glycidyl groups attached to the corners of the cage; characterized by the properties (e.g., epoxy equivalent weight 167; thermal stability (5 wt% loss) 365°C; and a liquid trisilanolisooctyl-POSS SO1455 (Hybrid, USA)); further marked as HO-POSS, an open cage POSS with seven isooctyl groups attached at the corners of the cage and three active silanol (hydroxyl groups) functionalities; and characterized by thermal stability (5 wt% loss) 363°C was applied as an additive at 3 wt% to the 50 ENR/PLA blend.

Additionally, the PLA formulations containing 3 wt% of Gly-POSS and HO-POSS, with properties described above were prepared.

Before the preparation of the PLA blends, the oven drying at 80°C during 24 h for polymer pellets was done. ENR/PLA blends were prepared by melt mixing in Brabender Lab Station Plasti-Corder, a small laboratory mixer with counter-rotating rotors. First the PLA pellets were placed in a mixing chamber and melted at 180°C. After 5 min the appropriate amount of epoxidized rubber was added. In the case of ENR/PLA/POSS blends, the appropriate amount (3 wt%) of Gly-POSS or HO-POSS, respectively, was added to PLA before the epoxidized rubber. The composition was mixed during 5 min at 150 rpm of rotor speed and then removed from mixer. After the mixing blends were again preheated at 180°C for 2 min and transferred to a mold, they are pressed at 200 bar during 10 min to form films. Similar procedure was applied to prepare the samples of PLA containing 3 wt% of various POSS molecules.

The tensile strength TS and the elongation at break of samples were measured according to ISO 37:2005 standard using universal testing machine Zwick Roell 1435 at room temperatures. The films were shaped into specimens by dumbbell-shaped cutter. The width and the length of the gauge region of the specimens were 4 mm and 25 mm, respectively. The speed of crosshead was 100 mm min^{-1}. The average values of six measurements were taken for each sample. The changes in mechanical properties after the degradation in soil during 1 month under controlled aging conditions (temperature 50°C, humidity 90%, pH = 5.5) were determined.

The dynamic viscoelastic properties of PLA/ENR blends at ambient temperatures were investigated using oscillation rheometer Ares G2 (TA Instruments) equipped with parallel-plate geometry (diameter of 25 mm). The oscillation amplitude sweep tests at temperature 25°C and constant angular frequency of 10 rad s^{-1} were conducted in oscillation strain range from 0.001 to 50%. Storage shear modulus G', loss shear modulus G'', and loss tan δ were analyzed. Studies of thermo-oxidative degradation were done using Ares G2 equipment, the samples were degraded at 180°C during 0, 30, 60, and 120 min under air atmosphere (flow rate 1 L min^{-1}), and then a frequency sweep tests at low oscillation strain (linear viscoelastic region) from 1 rad s^{-1} to 628 rad s^{-1} were performed.

The fitting of the storage shear modulus G' and the loss shear modulus G'' values to Maxwell's equations (4) and (5) and the calculation of the discrete relaxation spectrum were done by the TRIOS® Software provided by TA Instruments.

DSC analysis was performed using DSC1 apparatus (Mettler Toledo). All tests were performed under nitrogen atmosphere. The samples were subjected to two heating steps from 20 to 210°C, with a heating rate of 10°C/min. The objective of the first heating step was to eliminate the heat history of the sample.

The analysis of blend morphology was done using SEM microscope (JEOL Sendai Japan) for the fracture surface covered by gold. The ENR phase in 50 ENR/PLA samples was removed before the test by dissolving in toluene during 72 h.

5. Influence of Gly-POSS and HO-POSS on mechanical properties and crystallization of PLA/ENR blends: changes in mechanical performance of the blends after degradation in soil

The brittleness of PLA can be improved by elastomeric toughening additives. The polyisoprene rubber (IR), silicone rubber (SI), and acrylic rubber (AR) were used as a second component of the PLA blend to improve the mechanical performance of poly(lactic acid) [46]. Among all these blends, PLA/polyisoprene possessed high-tensile toughness due to its rubbery characteristic but low degree of crystallinity. Opposite the incorporation of acrylic rubber led to higher degree of crystallinity but quite low toughness. The toughness of PLA has been improved by the incorporating core-shell rubber particles [47]. It was found that the impact strength and the elongation at break increased without deterioration of thermal properties of final material. A matrix/dispersed morphology of various acrylonitrile content acrylonitrile-butadiene rubber/PLA blends was reported [48]. It was concluded that the incorporation of 10 wt% of rubber resulted in the enhancement of toughness and impact strength of neat PLA without any reduction in crystallinity. Desa et al. [49] investigated properties of PLA toughened by the addition of natural rubber (NR) and epoxidized natural rubber (ENR), after the addition of 5 wt% of rubber the impact strength of PLA increased from 3.35 kJ m^{-2} to 4.52 kJ m^{-2} (NR) and 4.14 kJ m^{-2} (ENR). The decrease of PLA crystallinity after incorporation of both rubbers was observed.

Since mechanical properties of immiscible polymer blends depend strongly on the formed morphology the analysis of SEM images was applied to observe the microstructure of the ENR/PLA. **Figures 1–4** display the changes in the microstructure of the blends. The EPM phase was dissolved and removed. The voids indicate the areas where the ENR phase was previously present. All blends showed phase separation confirming immiscibility of both phases. The droplets of ENR were dispersed throughout the PLA matrix. The PLA blend with 25 wt% of rubber characterized the smaller size of the elastomer domains. The increase of rubber loading up to 50 wt% resulted in higher melt strength during blend preparation at 180°C, retarding the rubber particle rupture during mixing consequently increasing the size of the rubber droplets in ENR/PLA blend. The elongated shape of rubber phase droplets was observed for the PLA blend containing 75 wt% of ENR. The addition of HO-POSS with isooctyl groups decreased the interfacial tension between the components of 50 ENR/PLA blend; as a consequence, the melt blending process was facilitated. The breakdown of rubber droplets was favorable leading to the better dispersion of rubber particles in PLA.

Dynamic mechanical characterization of obtained polymer blends at ambient temperature is able to give an additional insight onto the changes in mechanical performance after the incorporation of both POSS. Firstly, the attention of studies was focused on the effect of the various formulation of the ENR/PLA blend on the viscoelastic behavior; the changes of the storage shear modulus G′ as a function of oscillation strain at angular frequency of 10 rad s^{-1} were estimated (**Figure 5**). For the ENR/PLA composites, the storage shear modulus was found to be independent of strain up to low strain levels comparing to the neat uncured ENR rubber.

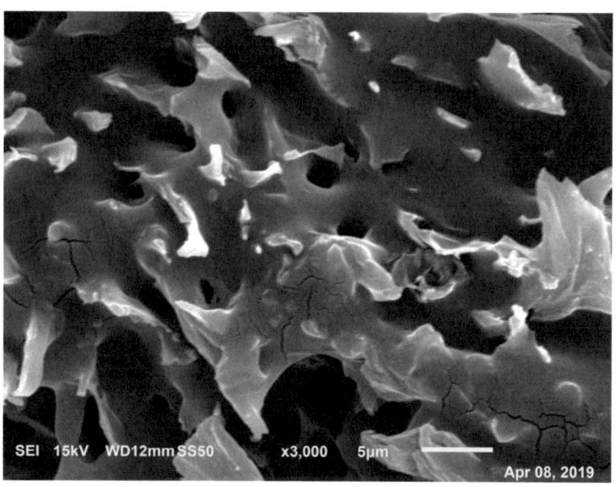

Figure 1.
SEM image of 25 ENR/PLA blend.

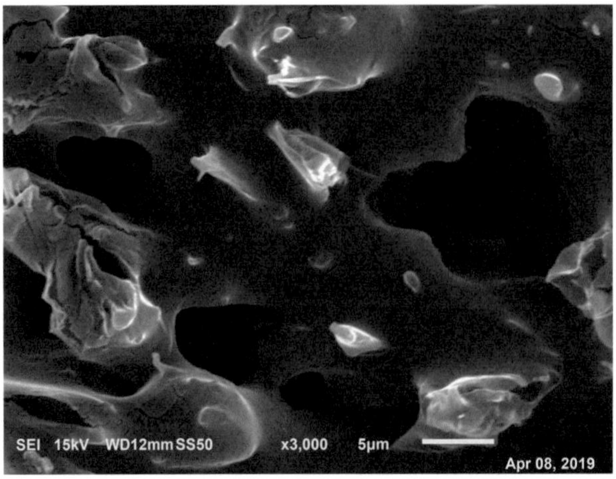

Figure 2.
SEM image of 50 ENR/PLA blend.

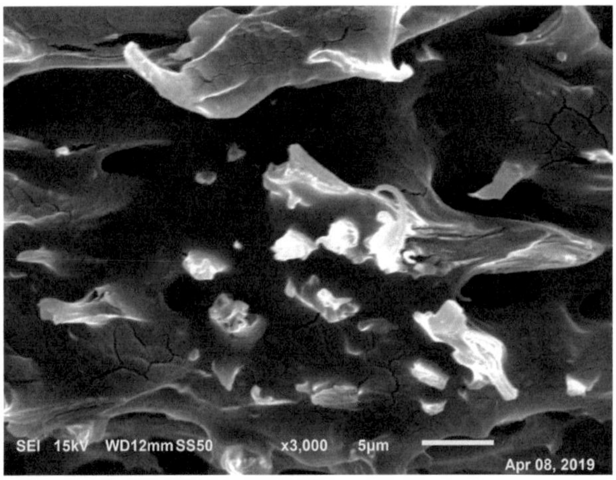

Figure 3.
SEM image of 50 ENR/PLA blend modified by HO-POSS.

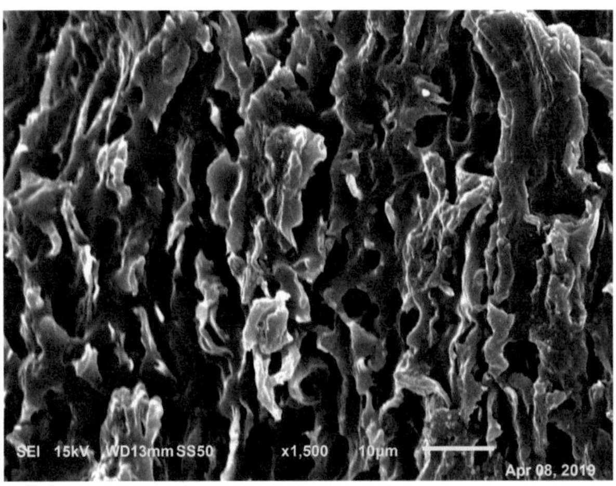

Figure 4.
SEM image of 75 ENR/PLA blend.

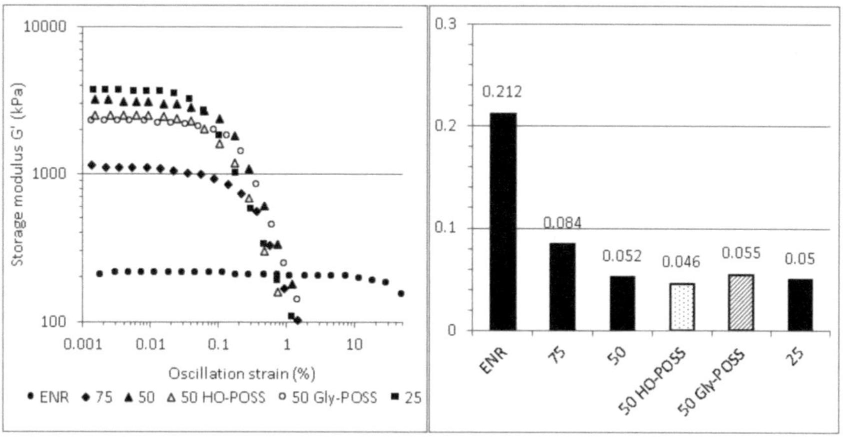

Figure 5.
Storage shear modulus G' as a function of oscillation strain (%) at 25°C for ENR/PLA blends with various ENR/PLA ratio (left). Damping properties and loss tan δ measured at linear viscoelastic region (oscillation strain of 0.01%, angular frequency of 10 rad s^{-1}, and temperature of 25°C) (right).

An increase of the oscillation strain above 0.2% caused strong decrease of storage shear modulus G′ for all formulations. As it was expected, the higher ratio of ENR in 75 ENR/PLA blend reduced significantly the values of storage shear modulus G′; at the same time, the loss tan δ increased, indicating stronger energy dissipation and better damping properties. It should be noted that both POSS molecules decreased the storage modulus G′ in comparison with neat 50 ENR/PLA blend. This effect can be attributed to a plasticization effect caused by the liquid Gly-POSS and HO-POSS, on the 50 ENR/PLA blend.

The more significant increase of the values of loss shear modulus as a function of oscillation strain was observed for 25 and 50 ENR/PLA blends (**Figure 6**). The higher applied force caused first the increase of the loss modulus G′ attributed to the higher dissipation of the energy, which resulted from the presence of the rubber phase in the blend, and then the microstructure of the blend collapsed leading to the decrease of both G′ and G″ moduli. Increasing the content of ENR in the blend caused the shift of the obtained maximum value of loss modulus G′ to a higher value of oscillation strain indicating that the formed microstructure was able to

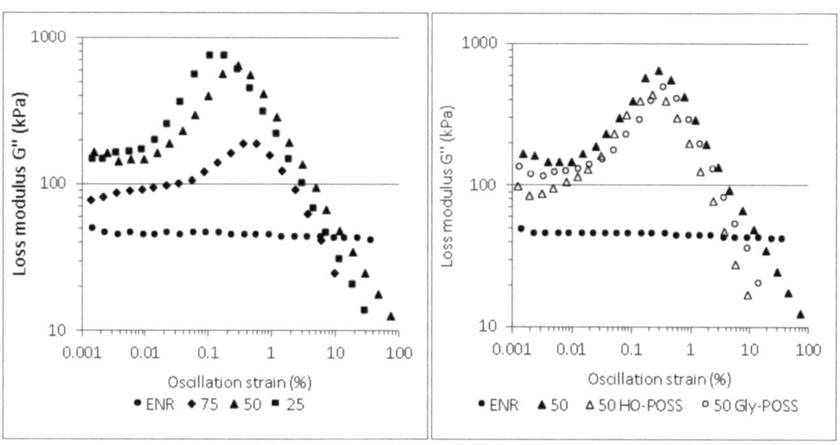

Figure 6.
Loss shear modulus G″ as a function of oscillation strain (%) at angular frequency of 10 rad s^{-1} and temperature of 25°C for ENR/PLA blends with various ENR/PLA ratio.

dissipate the energy up to higher level of deformation. For the composition containing POSS molecules, the lower values of loss shear modulus G′ were observed.

The mechanical properties of PLA blended with various content of ENR rubber are shown in **Table 1**. As it can be seen, the increasing ENR content led to the decrease of tensile strength, simultaneously increasing the elongation of break. Comparing all blends containing 50 wt% of ENR rubber, the tensile properties enhanced after the incorporation of both POSS additives. As the tensile properties of a polymer blends are correlated with the micromorphology, the better dispersion of rubber droplets in PLA matrix and the extent of the interfacial adhesion caused by the presence of functional groups in POSS molecules are responsible for the tensile strength increase observed for 50 ENR/PLA blend after the incorporation of both POSS.

In order to be assimilated by the microorganism present in soil, degradable polymer material needs to be fragmented. Even for the biodegradable materials such as poly(lactic acid), the unintentionally disposal of polymeric material could generate problems. It should be taken under consideration that soils can vary from place to place. The soil characteristic (temperature, water content, chemical composition, and pH) strongly influences on the degradation [50]. Degradation of polymer occurs due to the presence of oxygen, water, atmospheric pollutants, and agrochemicals. Activity of water, pH, temperature, and time must be controlled because they are factors promoting the hydrolytic degradation of polymers containing hydrolysable covalent bonds such as ester bonds. Moreover, the ratio of crystallinity affects the oxidative and hydrolytic degradation. The crystalline domains can restrict the diffusion of oxygen and water, limiting the chemical degradation [51]. The presence of rubber particles can be another factor strongly influencing on the degradation of the ENR/PLA blend. It was reported in literature that the degradation of polymers containing unsaturated bonds or backbone groups can occur via processes leading to crosslinking reactions or chain scission [52].

The changes in mechanical properties of the ENR/PLA blends were determined after the degradation of the material in soil during 1 month (**Table 1**). Additionally to estimate the effect of rubber on the degradation of the ENR/PLA blend, the aging coefficient K_C was calculated according to Eq. (1):

$$K_C = \frac{(TS * E_b)_{aged}}{(TS * E_b)_{befor\ ageing}}, \quad (1)$$

where TS is tensile strength and E_b is the elongation at break. The observed for the neat ENR value of the K_c coefficient (higher than one) and the increase of the tensile strength after aging confirmed that during composting, because of the

Composition of PLA/ENR	TS (MPa)	E_b (%)	TS^* (MPa)	E_b^* (%)	K_C
Neat ENR	0.50 ± 0.04	399 ± 31	0.68 ± 0.05	369 ± 73	1.26 ± 0.05
Neat PLA	24.7 ± 0.9	2.1 ± 0.5	3.43 ± 1.10	0.4 ± 0.4	0.03 ± 0.01
25 ENR/PLA	20.7 ± 3.6	3.8 ± 0.9	9.21 ± 1.2	2.7 ± 0.6	0.32 ± 0.05
50 ENR/PLA	10.4 ± 0.9	4.4 ± 1.4	6.99 ± 0.82	2.7 ± 0.7	0.41 ± 0.04
50 Gly-POSS	**12.0 ± 0.6**	**4.7 ± 1.6**	**3.88 ± 0.53**	**1.8 ± 0.9**	**0.12 ± 0.03**
50 HO-POSS	**11.0 ± 0.9**	**4.7 ± 1.4**	**3.60 ± 0.83**	**3.5 ± 0.7**	**0.24 ± 0.03**
75 ENR/PLA	6.9 ± 0.9	7.1 ± 2.4	5.68 ± 0.91	4.3 ± 1.1	0.50 ± 0.04

Bolded values are for the samples of blend containing POSS

Table 1.
Mechanical properties, tensile strength TS (MPa), and elongation at break E_b (%) of PLA/ENR blends with various PLA/ENR ratio. Tensile strength TS^* and elongation at break E_b^* after composting. Calculated values of compositing coefficient K_C.

Composition of blend	T_g °C	T_{cc} °C	T_m °C	χ_c
Neat PLA	55.4	106.1	156.8	2.7
25 ENR/PLA	55.3	124.9	148.5	1.9
50 ENR/PLA	55.3	128.3	147.3	0.8
50 Gly-POSS	49.4	—	—	—
50 HO-POSS	54.5	125.6	147.1	1.5
75 ENR/PLA	55.9	105.7	146.8	0.1

Table 2.
DSC analysis of ENR/PLA composites.

presence of unsaturated >C=C < bonds, the crosslinking occurred. In comparison with the neat PLA, the incorporation of rubber increased the values of K_C parameters, although lower than one values of the K_c parameter indicated that the partial degradation occurred. The changes in the microstructure of the blend caused by the degradation of PLA phase resulted in the deterioration of mechanical properties and in the decrease of the tensile strength TS and the elongation of break E_b of blend. The incorporation of Gly-POSS accelerated significantly the aging process, and lower value of K_c parameter was observed compared with 50 ENR/PLA blend. The decrease of K_c parameter resulted from the plasticizing influence of the liquid-like POSS on the PLA phase and the lower amount of crystalline phase. Both factors improved the diffusion of water and oxygen throughout the material, accelerating the degradation. Observations were confirmed by the degree of crystallinity calculated according to Eq. (2) [48]:

$$\chi_C = \frac{\Delta H_m - \Delta H_c}{w_{PLA} \Delta H_m^{o}}, \qquad (2)$$

where ΔH_m and ΔH_C are the enthalpies of melting and cold crystallization during the second heating cycle, respectively; $\Delta H^{o}{}_{m}$ is the enthalpy of 100% crystalline PLA homopolymer (93.7 J g^{-1}) [53]; and w_{PLA} is the weight fraction of PLA component in the blend (**Table 2**). The degree of crystallinity was calculated for the second cycle of heating, after erasing the thermal history of samples. During cooling no crystallization was observed. The crucial factor influencing the ratio of degradation in soil for 50 ENR/PLA blend modified by Gly-POSS in comparison with not modified 50 ENR/PLA blend and modified by HO-POSS was the amorphous structure and absence of crystalline phase. It should be also underlined that significant decrease in degree of crystallinity is observed for each ENR/PLA blend. The higher the amount of rubber, the lower the percentage of crystallization. Similar influence of epoxidized rubber on PLA was reported by Desa et al. [49].

6. Influence of POSS on the thermo-oxidative degradation of neat PLA and ENR/PLA blends

It is well known that the dynamic oscillatory measurements are very sensitive and allowed to characterize the changes occurred in the structure of melted polymers.

Different POSS were reported to be able to retard the changes of material properties occurring during processing; for this reason, the viscoelastic rheological characterization at 180°C was investigated to estimate the influence of POSS molecules on melt properties.

Figure 7 displays the evolution of complex viscosity as a function of angular frequency. It can be seen that η^* of the ENR/PLA blends containing higher rubber loading increased. Further, analyzing the influence of POSS incorporation the powerful influence of HO-POSS on η^* is observed for the whole frequency range in comparison with neat 50 ENR/PLA blend. The viscosity of POSS-polymer composite depends on the level of interactions. HO-POSS because of the presence of hydroxyl groups as well as isooctyl groups has higher affinity to both phases of the blend. Higher level of physical interactions between the polymer chains and HO-POSS resulted in increased viscosity of the melt in comparison with neat 50 ENR/PLA. This effect was not observed in case of Gly-POSS. As shown in **Figure 7** the thermo-oxidative degradation during 30 min led to the continuous increase of the complex viscosity of blends containing higher content of ENR rubber. The temperature probably caused the partial curing of the ENR phase. Additionally the formation of covalent bonds between hydroxyl groups present at the end of PLA chains and the epoxy groups of ENR was possible under these conditions of aging. Therefore all these reactions led to the changes in the polymer structure resulting in the formation of more branched structures and in the increase of the molecular weight. Taking into account that all degraded at 180°C samples showed during 30 min the increase of the complex viscosity η^* to quantify the effect of POSS for various degradation times, a relative degradation index RDI (3) was calculated according to the equation:

$$R.D.I = \frac{\eta_{time=0}}{\eta_{time}}, \quad (3)$$

where $\eta_{time=0}$ is the complex viscosity value at 1 rad s^{-1} for a sample not degraded and η_{time} is a the complex viscosity at 1 rad s^{-1} after 30, 60, and 120 min of thermo-oxidative degradation at 180°C. Hence, the parameter allows to quantify the increase in the melt viscosity due the crosslinking reactions or the formation of more branched structures with higher molecular weight (values lower than 1) or the drop of melt viscosity resulting from the chain scission (values higher than 1). The calculated values of RDI indexes for ENR/PLA POSS composites are compiled in **Figure 8**. The significantly lower values of RDI indexes observed for neat 50 ENR/PLA and Gly-POSS 50 ENR/PLA indicate that an increase of the melt viscosity occurred in higher extent than the HO-POSS 50 ENR/PLA blend. The increase in η^* at intermediate

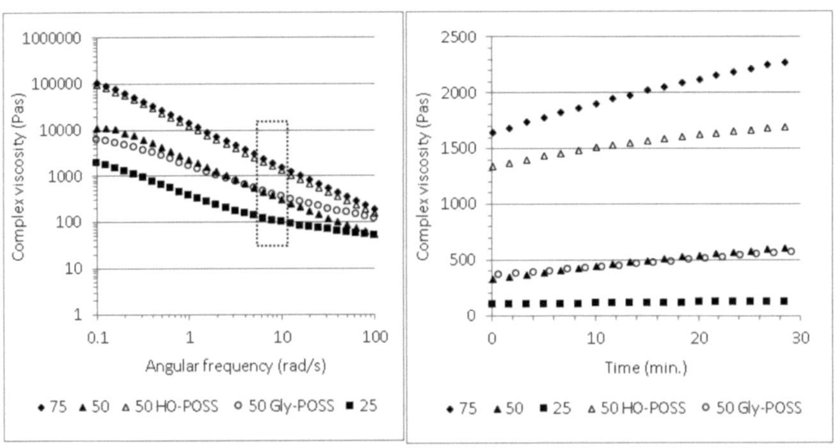

Figure 7.
Complex viscosity η^* as a function of angular frequency ω at 180°C (left). Complex viscosity as a function of time during 30 min of thermo-oxidative aging at 180°C (right).

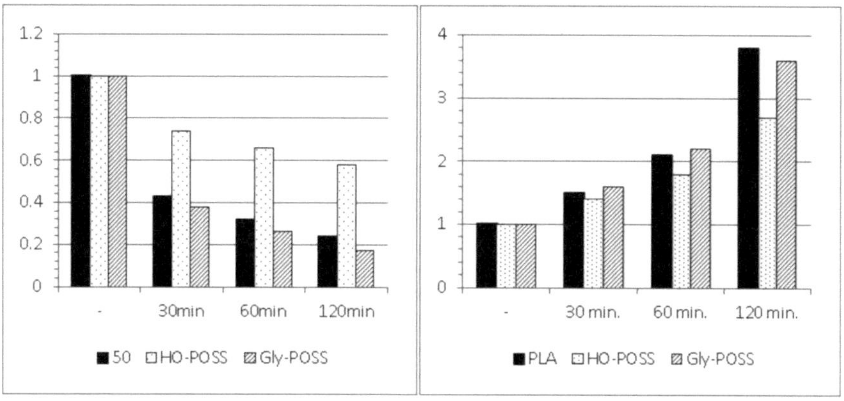

Figure 8.
Values of RDI indexes for the ENR/PLA blend (left) and PLA (right) modified by HO-POSS and Gly-POSS.

frequencies can relate to a restriction in chain mobility caused by chain branching or curing of ENR phase. During thermo-oxidative degradation of the melt blend, the total amount of the hydroxyl groups present at the end of the PLA chains increased, causing higher possibility to react with epoxy groups present in ENR. Comparing both POSS molecules, the weaker influence of HO-POSS on the increase of melt viscosity during thermo-oxidative degradation is observed. HO-POSS contains three silanol groups potentially able to react with epoxy functionalities of ENR, but these groups are covered by the isooctyl groups linked to the same silicon atom reducing the probability of the covalent bond formation with polymeric chains due to the steric hindrance. Moreover the higher melt viscosity of HO-POSS 50ENR/PLA blend (**Figure 7**) in comparison with neat 50 ENR/PLA formulation reduced the diffusion of oxygen preventing the degradation of both phases during processing.

Additionally **Figure 8** displays the evolution of RDI indexes for the neat PLA and PLA modified by 3 wt% of HO-POSS and Gly-POSS. All degraded samples presented similar behavior, which is typical for the thermo-oxidative degradation of PLA. The longer degradation time caused a decrease in melt viscosity, indicating the chain scission that was confirmed by the continuous increase of the RDI indexes after longer period of time. However, as for HO-POSS PLA, it can be indicated from **Figure 8** that the longer time of degradation favored the chain scission and decrease of melt viscosity, but the obtained viscosity reduction was restricted in comparison with neat PLA. Similarly, a reduction in values of RDI index after 120 min of thermo-oxidative degradation after incorporation of Gly-POSS is observed compared with neat PLA. Both POSS molecules enhanced the thermo-oxidative stability of neat PLA, although more significant stabilizing effect occurred for the HO-POSS. To analyze more deeply the influence of POSS molecules on the rheological behavior, the changes of storage modulus G′ and loss modulus G″ after thermo-oxidative aging during various times were determined. As an illustrative example, the evolution of both viscoelastic moduli as a function of frequency for 50 ENR/PLA blend is showed in **Figure 9**. As can be observed, all degraded samples of 50 ENR/PLA blend during various times of 30, 60, and 120 min showed predominant elastic behavior with storage modulus values G″ higher than loss modulus values G″ in whole studied frequency range. The thermo-oxidative degradation led to the increase of both moduli G′ and G″, and the increment of both moduli was dependent on the time of degradation. This fact pointed out that a degradation under this condition showed similar qualitative response, but the quantitative effect varied according to the time of degradation and was more accelerated during the first 30 min of degradation.

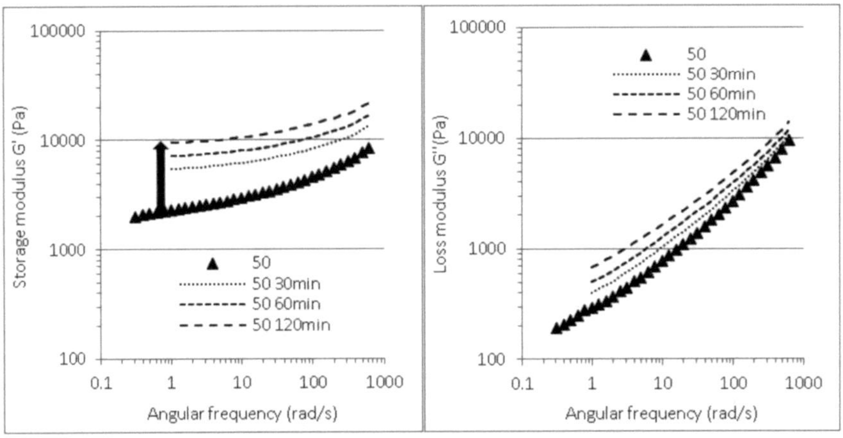

Figure 9.
Influence of the thermo-oxidative aging (30, 60, 120 min) on the viscoelastic properties (storage modulus G', loss modulus G") as a function of angular frequency at 180°C for 50 ENR/PLA blend.

For HO-POSS 50 ENR/PLA blend, when increasing degradation time, both moduli increased (**Figure 10**); regarding the increment of the G′ and G″ moduli, the presence of POSS molecules slowed the degradation of the blend. Interestingly, the opposite effect was observed for Gly-POSS 50 ENR/PLA; the degradation led to remarkable increase of the G′ modulus and decrease of the G″ modulus, indicating that the curing reactions occurred in higher extent. These results confirmed the observation from the analysis of melt viscosity and RMI indexes.

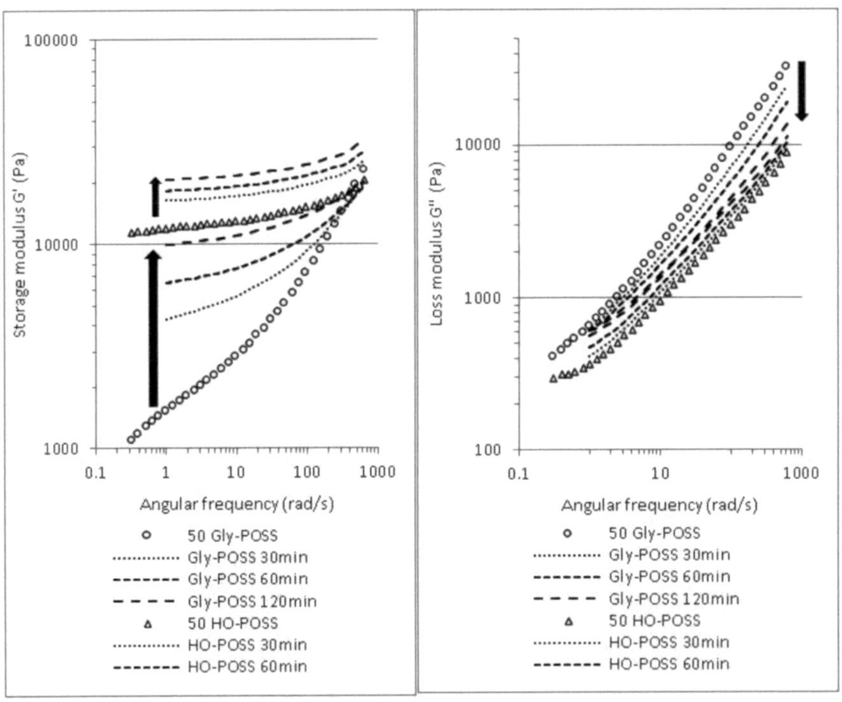

Figure 10.
Influence of the thermo-oxidative aging (30, 60, 120 min) on the viscoelastic properties (storage modulus G', loss modulus G") as a function of angular frequency ω for the POSS-modified PLA/ENR blends at 180°C.

The plasticizing effect of Gly-POSS on PLA was confirmed, and the addition of 3 wt% of Gly-POSS caused the decrease of the melt viscosity (not showed here); additionally lower values of storage modulus G′ were observed, indicating the decrease in chain entanglement and more free volume in the melt (**Figure 11**). For the PLA composites containing POSS molecules when increasing the degradation time, both moduli underwent a drop, and the effect was less predominant in the case of HO-POSS; thus, these investigations confirmed better thermal stability of HO-POSS formulation comparing with neat PLA.

The linear viscoelasticity data were used to obtain the discrete relaxation spectrum, which provides additional information about possible changes in PLA structure occurring during thermo-oxidative degradation. A discrete relaxation spectra is described by the generalized Maxwell model by the following relationships (4, 5) [54]:

$$G'(\omega) = \sum_{i=1}^{N} G_i \frac{(\omega \lambda_i)^2}{1 + (\omega \lambda_i)^2} \tag{4}$$

$$G''(\omega) = \sum_{i=1}^{N} G_i \frac{(\omega \lambda_i)}{1 + (\omega \lambda_i)^2} \tag{5}$$

where G_i and λ_i are the relaxation strength and relaxation time, respectively. It was found that for the experimental frequency range used in studies, the six Maxwell elements are sufficient for recalculation of the values of storage G′ and loss G″ moduli with high level of correlation ($R^2 = 0.999$). **Figure 12** presents the discrete relaxation spectrum for PLA not aged and degraded during 120 min samples. In **Table 3** calculated values of G_i and λ_i are compiled. The chain scissions and the decrease in the molecular weight of the polymer after 120 min of thermo-oxidative degradation were confirmed by the lower relaxation times. Finally, also from this study, the plasticizing effect of Gly-POSS on the behavior of not-aged PLA was observed, the enhanced mobility of PLA chains reflecting in the lower relaxation times as compared with PLA or HO-POSS PLA. The shorter relaxation time observed for neat PLA and Gly-POSS PLA degraded during 120 min, indicating the presence of the larger amount of the short PLA chains facilitating the relaxation of the molt.

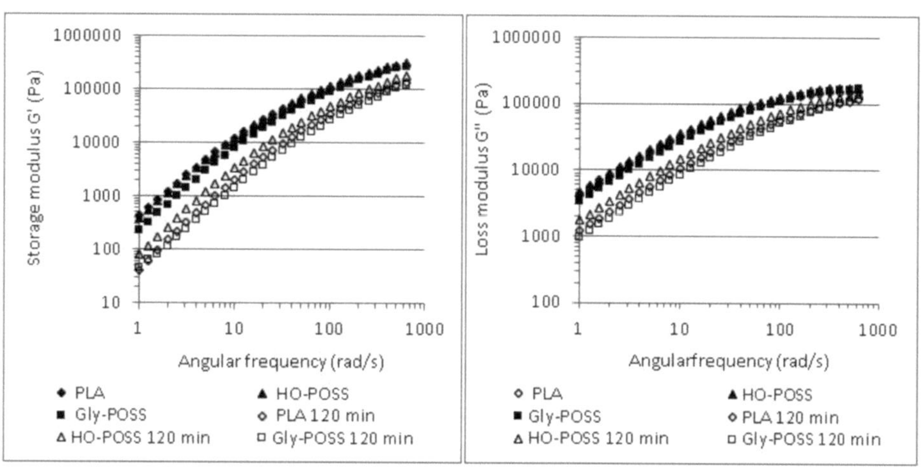

Figure 11.
Influence of the thermo-oxidative aging (120 min) on the viscoelastic properties (storage modulus G′, loss modulus G″) as a function of angular frequency ω for the POSS-modified PLA at 180°C.

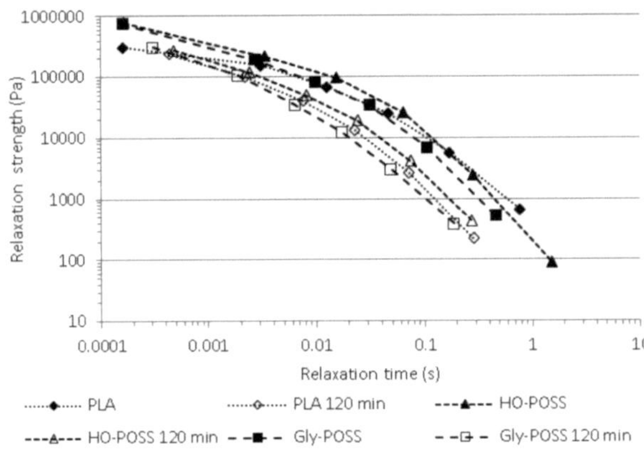

Figure 12.
Discrete relaxation spectrum for not aged and aged during 120 min PLA samples.

$R^2 = 0.9998$		Before aging		Time of aging 120 min	
Sample	**n**	**G_ikPa**	**λ_is**	**G_ikPa**	**λ_is**
PLA	1	308.888	$1.59 \cdot 10^{-4}$	235.202	$4.35 \cdot 10^{-4}$
	2	153.685	$3.06 \cdot 10^{-3}$	98.838	$2.23 \cdot 10^{-3}$
	3	67.539	0.01265	40.131	$7.53 \cdot 10^{-3}$
	4	24.786	0.04587	13.449	0.02266
	5	5.710	0.17028	2.662	0.07096
	6	0.665	0.76539	0.220	0.28426
PLA HO-POSS	1	783.398	$1.66 \cdot 10^{-4}$	275.002	$4.81 \cdot 10^{-4}$
	2	218.508	$3.31 \cdot 10^{-3}$	118.889	$2.45 \cdot 10^{-3}$
	3	95.374	0.01555	49.410	$8.13 \cdot 10^{-3}$
	4	25.367	0.06399	19.076	0.02409
	5	2.440	0.28063	4.138	0.07430
	6	0.090	1.50808	0.438	0.27293
PLA Gly-POSS	1	753.776	$1.59 \cdot 10^{-4}$	310.597	$3.06 \cdot 10^{-4}$
	2	191.331	$2.71 \cdot 10^{-3}$	106.000	$1.89 \cdot 10^{-3}$
	3	80.970	$9.87 \cdot 10^{-3}$	35.012	$6.27 \cdot 10^{-3}$
	4	33.950	0.03092	12.362	0.017141
	5	7.009	0.10387	3.062	0.04833
	6	0.521	0.46087	0.386	0.18548

Table 3.
The calculated values of relaxation strength G_i and relaxation time.

7. Conclusions

ENR/PLA composites were prepared by melt blending method. All prepared ENR/PLA composites showed immiscible droplet microstructure. The amount of added rubber strongly influenced on the tensile strength and elongation at break of blends. Incorporation of HO-POSS led to more homogenous morphology of the blend resulting in the enhancement of tensile strength as compared with 50 ENR/PLA blend. Both incorporated POSS molecules influenced on the dynamic performance of the 50 ENR/PLA blend, decreasing the values of the storage shear modulus. The effect was attributed to the lubricating influence of the liquid POSS molecules. The amorphous structure of the 50 ENR/PLA blend modified by Gly-POSS accelerated the degradation of the material in soil as compared to neat 50 ENR/PLA blend and HO-POSS-modified blend. The cold crystallization of the blend was affected by the presence of rubber phase, the degree of crystallinity meaningfully decreased after the addition of the higher rubber ratio. Dynamic oscillatory measurements in melt state indicated that an increase of the melt viscosity and the storage shear modulus occurred during thermo-oxidative degradation at various degradation times for ENR/PLA blends resulting from curing reactions. Significant increase of the storage modulus G′ during thermo-oxidative degradation of the 50 ENR/PLA blend modified by Gly-POSS was observed. The HO-POSS enhanced the thermal stability of blend resulted in lower changes of both modulus and melt viscosity values after thermo-oxidative degradation. Similar stabilizing effect of HO-POSS was observed for virgin PLA; the thermo-oxidative degradation of the melt and chain scissions occurred in the lower extent after the incorporation of HO-POSS. Considering that one of the disadvantages of PLA is its thermo-oxidative and thermomechanical degradation during processing at higher temperature, the enhanced thermo-oxidative stability after incorporation of HO-POSS is promising.

Author details

Magdalena Lipińska*, Klaudia Toczek and Magdalena Stefaniak
Chemistry Department, Institute of Polymer and Dye Technology, Lodz University of Technology, Lodz, Poland

*Address all correspondence to: magdalena.lipinska@p.lodz.pl

References

[1] Raftopoulos KN, Pielichowski K. Segmental dynamics in hybrid polymer/POSS nanomaterials. Progress in Polymer Science. 2016;**52**:136-187. DOI: 10.1016/j.progpolymsci.2015.01.003

[2] Zhang W, Camino G, Yang R. Polymer/polyhedral oligomeric silsesquioxanes (POSS) nano-composites: An overview of fire retardance. Progress in Polymer Science. 2017;**67**:77-125. DOI: 10.1016/j.progpolymsci.2016.09.011

[3] Zucchi IA, Galante MJ, Williams RJJ. Surface energies of linear and cross-linked polymers based on isobornyl methacrylate and methacryl-heptaisobutyl POSS. European Polymer Journal. 2009;**45**(2):325-331. DOI: 10.1016/j.eurpolymj.2008.10.025

[4] Goffin A, Duquesne E, Moins S, Alexandre M, Dubois P. New organic-inorganic nanohybrids via ring opening polymerization of (di)lactones initiated by functionalized polyhedral oligomeric silsesquioxanes. European Polymer Journal. 2007;**43**(10):4103-4113. DOI: 10.1016/j.eurpolymj.2007.07.041

[5] Chen JH, Chiou YD. Crystallization behavior and morphological development of isotactic polypropylene blended with nanostructured polyhedral oligomeric silsesquioxanes molecules. Journal of Polymer Science Part B: Polymer Physics. 2006;**44**(15):2122-2134. DOI: 10.1002/polb.20878

[6] Dintcheva NT, Morici E, Arrigo R, La Mantia FP, Malatesta V, Schwab JJ. Structure-properties relationships of polyhedral oligomeric silsesquioxane (POSS) filled PS nanocomposites. Express Polymer Letters. 2012;**6**(7):561-571. DOI: 10.3144/expresspolymlett.2012.59

[7] Zhou Q, Pramoda KP, Lee JM, Wang K, Loo LS. Role of interface in dispersion and surface energetics of polymer nanocomposites containing hydrophilic POSS and layered silicates. Journal of Colloid and Interface Science. 2011;**355**(1):222-230. DOI: 10.1016/j.jcis.2010.12.010

[8] Lewicki JP, Pielichowski K, Jancia M, Hebda E, Albo RLF, Maxwell RS. Degradative and morphological characterization of POSS modified nanohybrid polyurethane elastomers. Polymer Degradation and Stability. 2014;**104**:50-56. DOI: 10.1016/j.polymdegradstab. 2014.03.025

[9] Lipińska M, Soszka K. Viscoelastic behavior, curing and reinforcement mechanism of various silica and POSS filled methyl-vinyl polysiloxane MVQ rubber. Silicon. 2019;**11**:2293-2305. DOI: 10.1007/ s12633-019-0081-8

[10] Meng Y, Wei Z, Liu L, Liu L, Zhang L, Nishi T, et al. Significantly improving the thermal stability and dispersion morphology of polyhedral oligomeric silsesquioxanes/polysiloxane composites by in-situ grafting reactions. Polymer. 2013;**54**(12):3055-3064. DOI: 10.1016/j.polymer.2013.03.061

[11] Lipińska M, Imiela M. Morphology, rheology and curing of (ethylene-propylene elastomer/hydrogenate acrylonitrile-butadiene rubber) blends reinforced by POSS and organoclay. Polymer Testing. 2019;**75**:26-37. DOI: 10.1016/j.polymertesting.2019.01.020

[12] Konnola R, Nair CPR, Joseph K. Cross-linking of carboxyl-terminated nitrile rubber with polyhedral oligomeric silsesquioxane. Cure kinetics. Journal of Thermal Analysis and Calorimetry. 2016;**123**(2):1479-1489. DOI: 10.1007/s10973-015-5019-9

[13] Madhavan Nampoothiri K, Nair Nimisha R, John RP. And overview of the recent developments in

polylactide (PLA) research. Bioresource Technology. 2010;**101**(22):8493-8501. DOI: 10.1016/j.biortech.2010.05.092

[14] Tsuji H. Poly(lactide) stereo-complexes: Formation, structure, properties, degradation, and applications. Macromolecular Bioscience. 2005;**5**(7):569-597. DOI: 10.1002/mabi.200500062

[15] Srisa-ard M, Baimark Y. Effects of arm number and arm length on thermal properties of linear and star-shaped poly(D,L-lactide)s. Journal of Applied Science. 2010;**10**(17):1937-1943. DOI: 10.3923/jas.2010.1937.1943

[16] Martino VP, Ruseckaite RA, Jimenez A. Thermal and mechanical characterization of plasticized poly(L-lactide-*co*-D, L-lactide) films for packaging. Journal of Thermal Analysis and Calorimetry. 2006;**86**(3):707-712. DOI: 10.1007/s10973-006-7897-3

[17] Sodergard A, Stolt M. Properties of lactic acid based polymers and their correlation with composition. Progress in Polymer Science. 2002;**27**(6):1123-1163. DOI: 10.1016/S0079-6700(02)00012-6

[18] Middleton JC, Tipton AJ. Synthetic biodegradable polymers as orthopedic devices. Biomaterials. 2000;**21**(23):2335-2346. DOI: 10.1016/S0142-9612(00)00101-0

[19] Sun J, Yu H, Zhuang X, Chen X, Jing X. Crystallization behavior of asymmetric PLLA/PDLA blends. The Journal of Physical Chemistry. B. 2011;**115**(12):2864-2869. DOI: 10.1021/jp111894m

[20] Schmidt SC, Hillmyer MA. Polylactide stereocomplex crystallites as nucleating agents for isotactic polylactide. Journal of Polymer Science Part B: Polymer Physics. 2001;**39**(3):300-313

[21] Pistner H, Bendix DR, Muhling J, Reuther JF. Poly(L-lactide): A long-term degradation study in-vivo. Part III. Analytical characterization. Biomaterials. 1993;**14**(4):291-298. DOI: 10.1016/0142-9612(93)90121-H

[22] Qiu Z, Pan H. Preparation, crystallization and hydrolytic degradation of biodegradable poly(L-lactide)/polyhedral oligomeric silsesquioxanes nanocomposite. Composites Science and Technology. 2010;**70**:1089-1094. DOI: 10.1016/j.compscitech.2009.11.001

[23] Turan D, Sirin H, Ozkoc G. Effects of POSS particles on the mechanical, thermal, and morphological properties of PLA and plasticised PLA. Journal of Applied Polymer Science. 2011;**121**(2):1067-1075. DOI: 10.1002/app.33802

[24] Yilmaz S, Kodal M, Yilmaz T, Ozkoc G. Fracture toughness analysis of O-POSS/PLA composites assessed by essential work of fracture method. Composites Part B Engineering. 2014;**56**:527-535. DOI: 10.1016/j.compositesb.2013.08.081

[25] Wang R, Wang S, Zhang Y. Morphology, rheological behavior, and thermal stability of PLA/PBSA/POSS composites. Journal of Applied Polymer Science. 2009;**113**:3095-3102. DOI: 10.1002/app.30333

[26] Oshani BN, Davachi SM, Hejazi I, Seyfi J, Khonakdar HA, Abbaspourrad A. Enhanced compatibility of starch with poly(lactic acid) and poly(ε-caprolactone) by incorporation of POSS nanoparticles: Study on thermal properties. International Journal of Biological Macromolecules. 2019;**141**:578-584. DOI: 10.1016/j.ijbiomac.2019.09.026

[27] Kodal M, Wis AA, Ozkoc G. The mechanical, thermal and morphological properties of γ-irradiated PLA/TAIC

and PLA/OvPOSS. Radiation Physics and Chemistry. 2018;**153**:214-225. DOI: 10.1016/j.radphyschem.2018.10.018

[28] Kodal M, Sirin H, Ozkoc G. Effects of reactive and nonreactive POSS types on the mechanical, thermal, and morphological properties of plasticized poly(lactic acid). Polymer Engineering and Science. 2014;**54**(2):264-275. DOI: 10.1002/pen.23557

[29] Zubrowska A, Piotrowska E, Kowalewska A, Cichorek M. Novel blends of polylactide with ethylene-glycol derivatives of POSS. Colloid & Polymer Science. 2015;**293**(1):23-33. DOI: 10.1007/s00396-014-3344-3

[30] Tang L, Qiu Z. Effect of poly(ethylene glycol)-polyhedral oligomeric silsesquioxanes on the thermal and mechanical properties of biodegradable poly(L-lactide). Composites Communications. 2017;**3**:11-12. DOI: 10.1016/j.coco.2016.11.003

[31] Zhang X, Sun J, Fang S, Han X, Li Y, Zhang C. Thermal, crystalline, and mechanical properties of octa(3-chloropropylsilsesquioxane)/poly(L-lactic acid) hybrid films. Journal of Applied Polymer Science. 2011;**122**: 296-303. DOI: 10.1002/app.34059

[32] Huang L, Tan J, Li W, Zhou L, Liu Z, Luo B, et al. Functional polyhedral oligomeric silsesquioxane reinforced poly(lactic acid) nanocomposites for biomedical applications. Journal of the Mechanical Behavior of Biomedical Materials. 2019;**90**:604-614. DOI: 10.1016/j.jmbbm.2018.11.002

[33] Liu Z, Hu D, Huang L, Li W, Tian J, Lu L, et al. Simultaneous improvement in toughness, strength and biocompatibility of poly(lactic acid) with polyhedral oligomeric silsesquioxanes. Chemical Engineering Journal. 2018;**346**:649-661. DOI: 10.1016/j.cej.2018.03.077

[34] Monticelli O, Calabrese M, Gardella L, Fina A, Gioffredi E. Silsesquioxanes: Novel compatibilizing agents for tuning the microstructure and properties of PLA/PCL immiscible blends. European Polymer Journal. 2014;**58**:69-78. DOI: 10.1016/j.eurpolymj.2014.06.021

[35] Han D, Wen TJ, Han G, Deng YY, Deng Y, Zhang Q, et al. Synthesis of Janus POSS star polymer and exploring its compatibilization behavior for PLLA/PCL polymer blends. Polymer. 2018;**136**:84-91. DOI: 10.1016/j.polymer.2017.12.050

[36] Nijenhuis AJ, Grijpma DW, Pennings AJ. Highly crystalline as-polymerized poly(L-lactide). Polymer Bulletin. 1991;**26**:71-77. DOI: 10.1007/BF00299350

[37] Kalb B, Pennings AJ. General crystallization behaviour of poly(L-lactic acid). Polymer. 1980;**21**:607-612. DOI: 10.1016/0032-3861(80)90315-8

[38] Migliaresi C, de Lollis A, Fambri L, Cohn D. The effect of thermal history on the crystallinity of different molecular weight PLLA biodegradable polymers. Clinical Materials. 1991;**8**:111-118. DOI: 10.1016/0267-6605(91)90018-B

[39] Battegazzore D, Bocchini S, Frache A. Crystallization kinetics of poly(lactic acid)-talc composites. Express Polymer Letters. 2011;**5**(10):849-858. DOI: 10.3144/expresspolymlett.2011.84

[40] Tábi T, Sajó IE, Luyt AS, Kovács JG. Crystalline structure of annealed polylactic acid and its relation to processing. Express Polymer Letters. 2010;**4**:659-668. DOI: 10.3144/expresspolymlett.2010.80

[41] Vasanthakumari R, Pennings AJ. Crystallization kinetics of poly(L-lactic acid). Polymer. 1983;**24**:175-178. DOI: 10.1016/0032-3861(83)90129-5

[42] Kishore K, Vasanthakumari R, Pennings AJ. Isothermal melting behavior of poly(L-lactic acid). Journal of Polymer Science Polymer Physics Edition. 1984;**22**:537-542. DOI: 10.1002/pol.1984.180220401

[43] Yu J, Qiu J. Effect of low octavinyl-polyhedral oligomeric silsesquioxanes loadings on the melt crystallization and morphology of biodegradable poly(L-lactide). Thermochimica Acta. 2011;**519**:90-95. DOI: 10.1016/j.tea.2011.03.009

[44] Kopinke FD, Remmler M, Mackenzie K, Moder M, Wachsen O. Thermal decomposition of biodegradable polyesters-II. Poly(lactic acid). Polymer Degradation and Stability. 1996;**53**:329-342

[45] Zaharescu T, Rapa M, Lungulescu EM, Butoi N. Filler effect on the degradation of γ-processed PLA/vinyl POSS hybrid. Radiation Physics and Chemistry. 2018;**153**:188-197. DOI: 10.1016/j.radphyschem.2018.09.025

[46] Phattarateera S, Pattamaprom C. Comparative performance of functional rubbers on toughness and thermal property improvement of polylactic acid. Materials Today Communications. 2019;**19**:374-382. DOI: 10.1016/j.mtcomm.2019.02.012

[47] Lee JY, Chin IJ, Choi HJ. Effect of particle size and crosslinking on the toughening of core-shell type rubber modified poly(lactic acid) composites. Polymer Testing. 2018;**65**:440-449. DOI: 10.1016/j.polymertesting.2017.12.028

[48] Maroufkhani M, Katbab A, Liu W, Zhang J. Polylactide (PLA) and acrylonitrile butadiene rubber (NBR) blends: The effect of ACN content on morphology, compatibility and mechanical properties. Polymer. 2017;**115**:37-44. DOI: 10.1016/j.polymer.2017.03.025

[49] Desa MM, Hassan A, Arsad A, Mohammad N. Mechanical and thermal properties of rubber toughened poly(lactic acid). Advanced Materials Research. 2015;**1125**:222-226. DOI: 10.4028/www.scientific.net/AMR.1125.222

[50] Siracusa V. Microbial degradation of synthetic biopolymers waste. Polymers. 2019;**11**:1-18. DOI: 10.3390/polym11061066

[51] Lucas N, Bienaime C, Belloy C, Queneudec M, Silvestre F, Nava-Saucedo JE. Polymer biodegradation: Mechanism and estimation techniques. Chemosphere. 2008;**73**:429-442. DOI: 10.1016/j.chemosphere.2008.06.064

[52] Briassoulis D. The effect of tensile stress and agrochemicals Vapam on the ageing of low density polyethylene (LDPE) agricultural films. Part 1. Mechanical behavior. Polymer Degradation and Stability. 2005;**86**:489-503. DOI: 10.1016/j.polymdegradstab. 2004.11.021

[53] Liu H, Song W, Cheng F, Guo L, Zhang J. Interaction of microstructure and interfacial adhesion on impact performance of polylactide (PLA) ternary blends. Macromolecules. 2011;**44**:1513-1522. DOI: 10.1021/ma1026934

[54] Cuadri AA, Martin-Alfonso JE. Thermal, thermo-oxidative and thermomechanical degradation of PLA: A comparative study based on rheological, chemical and thermal properties. Polymer Degradation and Stability. 2018;**150**:37-45. DOI: 10.1016/j.polymdegradstab. 2018.02.011